The External Dimension of the European Union's Critical Infrastructure Protection Programme

External Dimension of the European Union's Critical Infrastructure Protection Programme: From Neighboring Frameworks to Transatlantic Cooperation provides the basis, methodological framework, and first comprehensive analysis of the current state of the external dimension European Programme for Critical Infrastructure Protection.

The challenges at the EU level are multidimensional insofar as identifying, designating and protecting critical infrastructures with the ultimate goal of harmonizing different national policies of the Member States and creating the identity of the European Union in this arena. Modern society has become so reliant on various sectors of critical infrastructure—energy, telecommunications, transport, finance, ICT, and public services—that any disruption may lead to serious failures that impact individuals, society, and the economy. The importance of critical infrastructures grows with the industrial development of global and national communities; their interdependence and resiliency is increasingly important given security threats including terrorism, natural disaster, climate change and pandemic outbreak

In the area of Critical Infrastructure Protection and Resilience, the European Union is constantly committed to setting the objectives for the Member States. At the same time, the European Commission promotes the importance of a common approach to Critical Infrastructure Protection (CIP) and ensures cooperation beyond the borders of the Union, while also cooperating with neighboring countries, including those soon willing to join the European Union.

This book has been structured and written to contribute to current critical infrastructures, resilience policy development and discussions about regional and international cooperation. It serves as a reference for those countries willing to initiate cooperation and that therefore demand deeper knowledge on the security cultures and frameworks of their potential partners.

Features:

- Provides an unprecedented analysis of the national frameworks of 14 neighboring countries of the EU, plus the United States and Canada
- Overcomes the language barriers to provide an overall picture of the state of play of the countries considered
- Outlines the shaping of national critical infrastructure protection frameworks to understanding the importance of service stability and continuity
- Presents guidelines to building a comprehensive and flexible normative framework
- Addresses the strategic and operational importance of international co-operation on critical infrastructure including efforts in CIP education and training
- Provides insight to institutions and decision-makers on existing policies and ways to improve the European security agenda

This book explains and advocates for establishing stronger, more resilient systems to preserve functionalities at the local, national, and international levels. Security, industry, and policy experts—both practitioners and policy decision-makers—looking for answers will find the solutions they seek within this book.

The External Dimension of the European Union's Critical Infrastructure Protection Programme

From Neighboring Frameworks to Transatlantic Cooperation

Edited by
Alessandro Lazari and Robert Mikac

CRC Press
Taylor & Francis Group
Boca Raton London

CRC Press is an imprint of the
Taylor & Francis Group, an **informa** business

First edition published 2022
by CRC Press
6000 Broken Sound Parkway NW, Suite 300, Boca Raton, FL 33487–2742

and by CRC Press
4 Park Square, Milton Park, Abingdon, Oxon, OX14 4RN

CRC Press is an imprint of Taylor & Francis Group, LLC

ISBN: 978-0-367-51718-2 (hbk)
ISBN: 978-1-032-22233-2 (pbk)
ISBN: 978-1-003-27376-9 (ebk)

DOI: 10.4324/9781003273769

Typeset in Garamond
by Apex CoVantage, LLC

Contents

PART FIVE OVERVIEW AND CONCLUSION

Abbreviations

AA/DCFTA	Association Agreement/Deep and Comprehensive Free Trade Area Agreement
AFAD	Turkish – T.C İçişleri Bakanlığı Afet ve Acil Durum Yönetimi Başkanlığı; Disaster & Emergency Management Authority
AM NREN CSIRT	Armenia National Research and Education Network Computer Security Incident Response Team
ATC	Anti-Terrorism Center
BiH	Bosnia and Herzegovina
BTC	Baku–Tbilisi–Ceyhan Pipeline
BTK	Turkish – Bilgi Teknolojileri ve İletişim Kurumu; Turkish Information and Communication Technologies Authority
CBRN	Chemical, Biological, Radiological, and Nuclear
CBSA	Canadian Border Services Agency
CCIRC	Canadian Cyber Incident Response Centre
CCRR	Canadian Cyber Resiliency Review
CCTX	Canadian Cyber Threat Exchange
CEI	critical energy infrastructure
CEPA	Comprehensive and Enhanced Partnership Agreement
CERT	computer emergency response team
CI	critical infrastructure
CII	critical information infrastructure
CIKR	critical infrastructure and key resources
CIMF	Cyber Incident Management Framework
CIMT	Critical Infrastructure Multimedia Tool
CIP	critical infrastructure protection
CIPR	critical infrastructure protection and resilience
CIS	Commonwealth of Independent States
CIRT	Critical Infrastructure Resilience Tool
CIS-IPA	Commonwealth of Independent States Inter-Parliamentary Assembly

CISA	Cybersecurity and Infrastructure Security Agency
CIWIN	Critical Infrastructure Warning Information Network
COST	European Cooperation in Science and Technology
CRRT	Cyber Rapid Response Teams
CSDP	Common Security and Defence Policy
CSIRT	Computer Security Incident Response Team
CSIS	Canadian Security Intelligence Service
CSS	Center for Strategic Studies
CSTO	Collective Security Treaty Organisation
CTED	The Counter-Terrorism Committee Executive Directorate
DCM	Decision of the Council of Ministers
DEA	Data Exchange Agency
DG ECHO	Directorate-General for Humanitarian Aid and Civil Protection of the European Commission
DHS	Department of Homeland Security
DCFTA	Deep and Comprehensive Free Trade Areas
EaP	Eastern Partnership
EAPC	Euro-Atlantic Partnership Council
EC	European Commission
ECI	European Critical Infrastructure
EEU	Eurasian Economic Union
EFDRR	European Forum on Disaster Risk Reduction
EIDHR	European Instrument for Democracy and Human Rights
ELEM	Elektrani na Severna Makedonija (Macedonian Electric Power Plants)
EM	Emergency Management
EMI	Emergency Management Institute
EMRA	Energy Market Regulatory Authority of Turkey
EnC	Energy Community
ENISA	European Network and Information Security Agency
ENP	European Neighbourhood Policy
EO	Executive Order
EPCG	Elektroprivreda of Montenegro (national electro-energetic company)
EPCIP	European Programme for Critical Infrastructure Protection
EPDK	Turkish – Enerji Piyasası Düzenleme Kurumu; Energy Market Regulatory Authority of Turkey
EPREV	Emergency Preparedness Review
ERE	Energy Regulatory Entity (Albania)

ERNCIP	European Reference Network for Critical Infrastructure Protection
EU	European Union
EUAM	EU Advisory Mission
EUBAM	EU Border Assistance Mission
FEMA	Federal Emergency Management Agency
FPRS	Fire Protection and Rescue Service (Albania)
FPT	Federal Provincial and Territorial
FS-ISAC	Financial Services – Information Sharing and Analysis Center
GAO	Government Accountability Office
GCI	Global Competitiveness Index
GDP	Gross domestic product
GDPR	General Data Protection Regulation
GIS	geographic information system
HSEM	Homeland Security and Emergency Management
IAC	Interstate Aviation Committee
IAEA	International Atomic Energy Agency
IBM	International Business Machines
IBRD	International Bank for Reconstruction and Development
ICDO	International Civil Defence Organisation
ICS	industrial control systems
ICT	information and communication technology
IICC	Infrastructure Interagency Coordinating Council
IN	infrastructure networks
IPA	The Instrument for Pre-Accession Assistance
IPAP	Individual Partnership Action Plans (NATO)
ISAC	Information Sharing and Analysis Center
ISACA	Information Systems Audit and Control Association
ISO	International Organization for Standardization
ISP	internet service provider
ISS	Intelligence and Security Service
IT	information technology
ITU GCI	International Telecommunication Union's Global Cyber Security Index
JANAF	Jadranski naftovod (Adria Pipelines)
JSC	joint-stock company
JSC MEPSO	Joint Stock Company for electricity transmission and power system control
JSC OKTA	Oil Refinery Skopje
KAPS	Kosovo Academy for Public Safety

KESH	Korporata Elektroenergjitike Shqiptare (The Energy of Albania)
LEPL	legal entity of public law
LFD	Lead Federal Departments
MECI	Ministry of energy and coal industry of Ukraine
MIA	Ministry of Internal Affairs
MIE	Ministry of Infrastructure and Energy (Albania)
MoI	Ministry of Interior
NAC	National Anti-Corruption Centre
NAECCS	National Authority for Electronic Certification and Cyber Security (Albania)
NATO	North Atlantic Treaty Organization
NCCIC	National Cybersecurity and Communications Integration Center
NCPA	National Civil Protection Agency (Albania)
NCSF	National Cross Sector Forum
NDS	National Defense Strategy
NERC	North American Electric Reliability Corporation
NGO	non-governmental organization
NICS	Next-Generation Incident Command System
NIPP	National Infrastructure Protection Plan
NIS	network and information system
NISS	National Institute for Strategic Studies of Ukraine
NISS	National Critical Infrastructure System of Azerbaijan
NITP	National Inspectorate of Territorial (Albania)
NFES	National Fund for Entrepreneurship Support
NNCEIP	Non-Nuclear Critical Energy Infrastructure Protection
NPRD	National Protection and Rescue Directorate
NSS	National Security Strategy
NSS	National Security Service (Armenia)
OAC	Operational and Analytical Center
OECD	Organisation for Economic Co-operation and Development
OSP	Operator Security Plans
OPCW	Organisation for Prohibition of Chemical Weapons
OSCE	Organization for Security and Co-operation in Europe
OT	operational technologies
PA CSTO	Parliamentary Assembly of the Collective Security Treaty Organisation
PCA	Partnership and Cooperation Agreement
PECB	Professional Evaluation and Certification Board
PESCO	Permanent Structured Cooperation

PCCIP	US President's Commission on Critical Infrastructure Protection
PDP	Professional Development Programme
PfP	Partnership for Peace
PPD	Presidential Policy Directive
PPP	public private partnership
PPRD	prevention, preparedness and response to natural and man-made disasters
RA	Republic of Armenia
RACVIAC	Centre for Security Cooperation (formerly Regional Arms Control Verification and Implementation Assistance Center)
R&D	research and development
RCMP	Royal Canadian Mounted Police
RECIPE	Resilience of Critical Infrastructure Protection in Europe (EU funded project)
RRAP	Regional Resilience Assessment Program
RS	Republic of Serbia
RS	Republika Srpska (Bosnia and Herzegovina entity)
SAA	Stabilization and Association Agreement
SCADA	Supervisory Control and Data Acquisition
SEE	South-Eastern Europe
SEECSC	South East Europe Cyber Security Center
SME	Small- and medium-sized enterprise
SOME	Turkish – Sektörel Siber Olaylara Müdahale Ekipleri; sectoral CERT
SSA	sector-specific agency
SSSCIP	State Service of Special Communication and Information Protection of Ukraine
SSSG	State Security Service of Georgia
SSM	Turkish – Savunma Sanayii Müsteşarlığı; Undersecretariat for Defence Industries
STII	State Technical and Industrial Inspectorate (Albania)
STM	Turkish – Savunma Teknolojileri ve Mühendislik Ticaret A.Ş; Defense Technologies Engineering and Trade Inc
TAF	Turkish Armed Forces
TAIEX	Technical Assistance and Information Exchange instrument of the European Commission
TANAP	Trans-Anatolian Natural Gas Pipeline
TAP	Trans Adriatic Pipeline
TeC	Transport Community

TRCC	The Resiliency Council of Canada
TTX	table-top exercise
TÜBİTAK	Turkish – Türkiye Bilimsel ve Teknolojik Araştırma Kurumu; The Scientific and Technological Research Council of Turkey
UCP	University of Civil Protection (Belarus)
UK	United Kingdom
UN	United Nations
UNOCT	United Nations Office of Counter-Terrorism
UPTC	Urban Protection Training Center Montenegro
US	United States
USA	United States of America
USAID	United States Agency for International Development
USOM	Turkish – Ulusal Siber Olaylara Müdahale Merkezi; national Computer Emergency Response Team (CERT)
VRAC	Virtual Risk Analysis Cell
WBIF	Western Balkans Investment Framework Initiative

Figures

Tables

Acknowledgments

The editors would like to thank and personally acknowledge the unprecedented efforts undertaken by the community of experts that have provided their time and availability for the accomplishment of this publication. Without their commitment and willingness to share the state of the art of their respective countries, this publication would not have seen the light.

The editors would also like to express their gratitude to Ivana Cesarec for her special effort and dedication in supporting the conceptualization, drafting, review and finalization of this manuscript.

Finally, the editors would like to thank the community of the EPCIP and ERNCIP's points of contact, members and experts with whom they have had the privilege to work all over these years. Within this wonderful context, made of motivated and open-minded colleagues and friends, the spark that has brought this book to life has been able to ignite.

Alessandro Lazari and Robert Mikac, June 2021

Foreword

Critical infrastructure protection and resilience are becoming an emerging challenge for the European Union. The disruption of such infrastructure through physical and cyber-attacks has a cascading, cross-sectoral and cross-country effect. The recent ransomware cyber-attack of the US Colonial Pipeline is an example of the vulnerability of critical infrastructures. It also shows how such cyber-operations can severely affect societal security in its entirety, endangering access to energy or even endangering health. Therefore, the evolving nature of threats against European critical infrastructure (ECI) makes it necessary to review and assess the state-of-play of ECI and its resilience capability toward man-made and natural disasters.

For the European Union, the 2019 review of the Council Directive on the identification of European critical infrastructures and the assessment of the need to improve their protection is an important step in creating a common approach regarding the identification of and providing needs to protect critical European infrastructure. The Commission's new legislative proposal of 2020 aims to improve the Directive's scope, by introducing Critical Entities Directive (CER) which includes 10 distinct sectors, and align it with the revision of the EU Network and Security Information Directive (NIS). This also ties into the EU's Security Union Strategy 2020–2025 with the aim of protecting and enhancing the physical and cyber-resilience of critical infrastructures.

While these measures go a long way in increasing critical infrastructure resilience in the EU, we recognize that a chain is only as strong as its weakest link and that these links extend beyond the boundaries of the Union. As part of the Technical Assistance and Exchange Instrument (TAIEX), the EU initiated a dozen of programs to strengthen its neighborhood, organizing expert missions and study visits. These programs are administered by DG NEAR and supported by DG HOME and DG ECHO. Through the EU Chemical, Biological, Radiological and Nuclear (CBRN) Centers of Excellence, protection of critical infrastructure EU projects are also developing with their direct neighbors.

Regarding the emerging field of critical infrastructure protection, challenges remain in identifying best practices for the EU and its partners. Therefore, the intention of this book is to present the current state of developments of strengthening the resilience of critical infrastructure at the EU level as well as its wider neighbourhood such as EU enlargement countries and the Eastern Partnership. The findings also intend to seek out new areas of cooperation with transatlantic partners such as the United States and Canada. Experts will find the multi-level analysis useful when sharing knowledge and best practices in the field. The interdisciplinary scope increases the relevance of the book, not only for public and private actors but for academia and the general public as well. I would like to thank the authors, not only for providing the reader with knowledge of EU policy cooperation and development, but also for providing a wide-ranging insight into different CIP perspectives and providing a complex analysis to raise awareness and common knowledge of critical infrastructure security across and beyond Europe.

Director General Maciej Popowski
Directorate-General for
Neighbourhood and Enlargement Negotiations (DG NEAR)
European Commission

Editor Biographies

Alessandro Lazari is Senior Key Account Manager at F24 in Germany. He has worked as a specialist on critical infrastructure protection, resilience and cybersecurity since 2004. He holds a Master's degree in Law (University of Bologna) and a Ph.D. in computer engineering, multimedia and telecommunications (University of Florence). Between 2010 and 2019, in his role of Scientific Officer for the European Commission' Joint Research Centre, he worked on policy support to the "European Programme for Critical Infrastructure Protection (EPCIP)," to "Strengthening Europe's Cyber Resilience System and Fostering a Competitive and Innovative Cybersecurity Industry." He has also contributed to some of the initiatives related to the "external dimension of the EPCIP."

Alessandro is a fellow in Legal Informatics at the School of Law of the University of Lecce, Italy, a lecturer on the course "Protecting Critical Infrastructures" at NATO's Centre of Excellence Defence Against Terrorism (COEDAT) and co-founded the Center for Interdisciplinary Research on Critical Infrastructure Security and Resilience (CRISR) at the University of Salento, Italy. He is the author of the book European Critical Infrastructure Protection (2014).

Robert Mikac is Associate Professor at the Faculty of Political Science of the University of Zagreb in the area of social sciences, the field of political science, and the subfield international relations and national security. The areas of his interest and expertise are international relations, international and national security, security management, crisis and disaster management, civil protection, Afghanistan, privatization of security, critical infrastructure protection and resilience, and migrations and security. He is the author and co-author of six books and more than 40 scientific articles. At his previous workplace in National Protection and Rescue Directorate, he was in charge of affairs related to critical infrastructure, and from 2012 till 2015 the national point of contact for critical infrastructure. Prior to his academic career, he served in various positions within the Croatian security sector from the operational to the strategic level in the country and internationally. Among other things, he served in the NATO ISAF mission in Afghanistan. For his merit in Afghanistan, he was awarded the NATO Medal and the Order of Merit of the Italian Republic.

Contributor Biographies

Ivana Cesarec works for the Ministry of Interior, Republic of Croatia (Civil Protection Directorate) as Head of the Department of Critical Infrastructure and Cultural Heritage. She is also a European critical infrastructure protection point of contact of the Republic of Croatia. As an assistant lecturer at the University of Applied Sciences, Velika Gorica (Croatia), she teaches two courses: critical infrastructure and critical infrastructure management. From 2013 to 2018, she was an employee of the Department for Prevention, Planning and Analytics within the Civil Protection Sector, National Protection and Rescue Directorate of Republic of Croatia. She holds a Master's degree in International Relations and Diplomacy and a Professional Specialist Engineer of Crisis Management and Logistics degree. She is an author and co-author of several scientific and professional papers and has participated in various national and international training programs.

Salih Bıçakcı is Associate Professor at the International Relations Department of the Kadir Has University, Istanbul. His research primarily focuses on security, cyber security, critical infrastructure protection, hybrid security, data privacy and terrorism. He has made evaluations and presentations on cyber security at the NATO Defense Against Terrorism Centre of Excellence (COEDAT), NATO Command and Control Centre of Excellence (C2COE) and NATO Maritime Security Centre of Excellence (MARSEC). He also taught Cyber Security and Middle Eastern Security courses at the Armed Forces Academy of the Turkish War College.

Ayhan Gücüyener Evren is a Ph.D. candidate at the international relations department of the Kadir Has University, Istanbul. Her research interests are cyber security in international relations, critical infrastructure protection and resilience policies.

Mirza Smajić is Associate Professor at the Faculty of Political Science in Sarajevo. His research interests include security studies, human security, national security and law enforcement and police studies. As an author or co-author, he published four books and studies, and a number of professional and scholarly papers published in BiH and abroad. He participated as a researcher or consultant in a large number of congresses and projects in BiH, Serbia and Croatia.

Zoran Keković is Professor at the Faculty of Security Studies, Belgrade University, Serbia, where he established and headed the first academic postgraduate program in the field of crisis management in Southeastern Europe. He is the founder and president of the Center for Risk Analysis and Crisis Management. He is the organizer and participant of many scientific and professional conferences on the topic of security, as well as the author of a number of papers, monographs and textbooks in the field of security studies, corporate security, risk management and crisis management.

Olivera Injac is Associate Professor at the Department of Security Studies – Faculty of Humanistic Studies of the University of Donja Gorica, Montenegro, and the Minister of Defense of Montenegro. She started to work at the university in 2008, and at previous workplaces in the Ministry of Interior of Montenegro, she worked on the analysis, advising and international affairs positions. She was Chair of the Department of Security Studies from 2011 to 2016, at the Faculty of Humanistic Studies. Olivera Injac is a full-time faculty member and teaches on the subjects of international security, security strategies, global security, international terrorism and security culture. She is the author of two books, a few book chapters and about 30 scientific papers and expert analysis. Her research interests and expertise include security studies, international terrorism, global risks and national security.

Marina Mitrevska is Professor at the Institute for Security, Defence and Peace at the Faculty of Philosophy, University of Ss. Cyril and Methodious in Skopje, Republic of North Macedonia. She is the Head of the third cycle of doctoral studies in security, defense and peace. She is a member of the Accreditation and Evaluation Board of Higher Education in the Republic of North Macedonia. She is Editor-in-Chief of the international scientific journal *Contemporary Macedonian Defence*. Her field of scientific research is security, diplomacy, peacekeeping operations and crisis management. She is actively engaged in researching and publishing scientific papers and books in the field of security. She is the author of 11 books and more than a hundred scientific papers.

Stavri Dhima is a freelance senior energy expert with a wide and deep experience and knowledge in several energy sectors, as well as a very long and successful experience in scientific and administrative engagement in upstream, midstream and downstream petroleum activities and particularly in the gas sector. He graduated from the University of Tirana with very high achievements and evaluation in physics during 1974–1975. In 1997, he finished his post-university studies and earned his doctorate in geophysics. In 2001, he became Associate Professor. During the

1979–1999 period, he had been engaged in Geophysical Enterprise and on Oil and Gas Institute in Fieri, finally as Director of this Institute. From 1999 up to the end of 2019, when he retired, he had been engaged on several positions in the high administration of the Albanian Ministry responsible for Energy and Infrastructure, as General Director of the General Petroleum Directorate, and as well as the Head of sector of energy policies and projects on energy sector, mainly in the oil and gas activities. During the time that he has been engaged in several positions in the high administration, he gave a distinguished contribution on the completion of the legal and institutional framework for the Albanian petroleum and energy sectors through the transposition of the European legal and institutional framework, and for the preparation of the strategic document for the development of the energy sector in Albania, especially for the energy infrastructure. His research and work in scientific and technical areas are geological and geophysical studies for petroleum exploration and development, studies on the management, administration and operation of petroleum and energy sector in Albania, and national and regional projects on petroleum and energy infrastructure.

Ardi Dhima, MSc, Urbanist/Architect, is a freelance infrastructure expert, architect and urban planner in the energy sector. He has experience of more than 10 years, in the oil and gas industry, as a project and permit manager, for the infrastructure, development, and construction of several oil and gas–related facilities (pipelines, deposits, oil production and exploration wells, warehouses, etc.). During recent years he has taken part in several training programs on efficient energy use and planning (Life Academy, Sweden, 2014), oil and gas operational safety (SETA Training and Advisory Services Ltd., Tirana, 2016).

Sara Prenga recently graduated from the Faculty of Political Science, University of Zagreb, majoring in national security. She is a native speaker of Albanian and Croatian, and additionally actively speaks several world languages.

Dardan Koçani is Director of the Department of European Integration and Policy Integration at the Ministry of Internal Affairs of the Republic of Kosovo. He holds an MA in European Studies from the European College of Parma and University of Parma and a BS with a major in management, economics and statistics and a minor in legal studies from the Rochester Institute of Technology. Mr. Kocani has been working at the Ministry of Internal Affairs since 2011, where he held other positions such as the Head of Division for Policy Coordination and Senior Official for Policy Analysis. He has finished the Civil-Military Response to Terrorism course at the Center for Civil-Military Relations/Naval Postgraduate School in the US, an 11-week program on Advanced Security Studies – Capacity Building, and other programs at Geneva Centre for Security Policy, the George. C. Marshall Center for Security Studies, and Swiss Federal Institute of Technology.

Oleksandr Sukhodolia is Professor and Head of Energy Security and Technogenic Safety at the National Institute for Strategic Studies, Ukraine. He has over 20 years of scientific research and educational activity as well as and public service experience in the field of energy security. His research interests focus on national and energy securities, energy policy, and critical infrastructure protection. He is a member of government working groups on the development of critical infrastructure protection concepts in Ukraine.

Nadya Karpenko graduated from the Belarusian State Economic University (Minsk, Belarus) and earned the title of Master of Intercultural Communications. She then graduated in Political Science at the University of Vienna while currently studying at the Central European University (Vienna, Austria). Regarding her professional experience, she worked as a trading platform administrator for United Nations Development Program (UNDP) in Belarus from 2014 till 2015. She was the freelance events coordinator for the Academic Council of the United Nations System (Vienna, Austria) from 2018 till 2020. At the same time, her last position was from 2019 to 2020 as a junior research assistant at the Institute for Peace Support and Conflict Management at the National Defense Academy (Vienna, Austria).

Denis Cenusa is a researcher in the DFG project "Domestic Preferences and EU Cooperation. Explaining Foreign Policies in Ukraine, Georgia and Moldova: A Two-Level-Game Analysis" at the Institut für Politikwissenschaft, Justus-Liebig-Universität (Germany). At the same university, he conducts his Ph.D. work that examines the EU external governance impact on state resilience in Eastern Europe based on the cases of Georgia, Moldova and Ukraine. He obtained a Master's diploma in European Interdisciplinary Studies at the College of Europe Natolin (2013) and in Political Sciences at the State University of Moldova (2007). In 2014, he also graduated from the Advanced Program on EU Law and Economics at Riga Graduate School of Law. He is affiliated with a Think-Tank "Expert-Grup" in Moldova since 2013, where he coordinates the Energy Security Department. In this capacity, he contributed to the drafting of the 2030 National Development Strategy of Moldova. Between 2015 and 2020, he was the national coordinator of the project "Understanding the Association Agreements between the EU and Ukraine, Moldova and Georgia", implemented by the Center for European Policy Studies (Brussels) and financed by the Swedish government. His fields of expertise comprise good governance in post-Soviet space, EU external governance in Eastern Europe and Russia, European energy and visa policies, and state resilience.

Grigol Modebadze is Head of the Socio-Economic and Energy Security Division at the Office of the National Security Council of Georgia. Prior to his current position, he worked at State Security Service as an economic security expert (2018–2019).

Starting from February 2014 till September 2017, he worked as a senior advisor, and from September 2017 till January 2018, he worked as an acting director of the economic security department at the Office of State Security and Crisis Management Council of Georgia. Mr. Modebadze worked as an economist at the World Bank Caucasus Office (2011–2013). Also, he held the senior researcher position at the International School of Economics at Tbilisi State University (TSU) (ISET) during 2008–2011. Additionally, Mr. Modebadze has worked on a range of projects in Georgia as a local expert for the World Bank, USAID, UNDP, NATO and other international organizations during his career. Furthermore, he lectured at the Ivane Javakhishvili Tbilisi State University during 2008–2011 and since September 2011 he is lecturing at the Agrarian University of Georgia. Mr. Grigol Modebadze studied Finance and Banking at Ivane Javakhishvili Tbilisi State University. He holds a Master's degree in Economics from the International School of Economics at TSU (ISET) and he is a Ph.D. candidate at TSU. Mr. Modebadze is managing a team of senior advisors, who analyze and report an issue of broad scope and complexity pertaining to State Security within the field of Economy, Energy, Environment and Social affairs. Also, he is a contact person for the Critical Infrastructure Protection reform in Georgia.

Ashkhen Beglaryan, Ph.D. in Political Science (2017 National Defense Research University, MoD, Republic of Armenia), is Senior Lecturer at the Department of Creative Industries at the Russian-Armenian University (Yerevan, Armenia). Mrs. Beglaryan gives lectures at Russian-Armenian University on political communication. Prior to that, she was a lecturer at the Bryusov State University on the political discourse and analysis of political processes and at French Lycee in Armenia on the essentials of sociology and political science. She has MA and BA degrees in sociology. Mrs. Beglaryan did a Ph.D. in the Department of Political Sciences on "The civil society within the context of national security of the Republic of Armenia". She is the author of six scientific articles on different issues of the national security of the Republic of Armenia. Mrs. Ashkhen Beglaryan is a member of staff of the National Security Council of the Republic of Armenia since 2008.

Alikhayyam Guluzade is the Head of the cyber defense and monitoring department at PROSOL CJSC (Azerbaijan Technology Power House). Prior to joining PROSOL, Alikhayyam worked for KPMG, Deloitte, Central Bank of Azerbaijan Republic. He has more than eight years of experience in IT and information security risk governance and management. His project portfolio contains tens of IT and information security–related projects in Azerbaijan and also in the CIS region. He was also a lecturer in Financial Mathematics and Management of Information Systems at Baku State University. Alikhayyam Guluzade has practiced in several industries, as well as in finance, telecommunication, manufacturing, hospitality, oil and gas, aviation and so on.

Ayshan Suleymanova is a cyber security incident analyst at PROSOL CJSC (Azerbaijan Technology Power House). She graduated from Azerbaijan State Oil and Industrial University. Currently, she is studying for Master of Science in Software Engineering at the University of Tartu. She has experience in multiple projects in Azerbaijan in the development and improvement of security incidents management optimization. She also participated and led several startups in Eastern Europe and the Caucasus region.

Matthew Vatter is a retired senior army officer from the Minnesota National Guard. During his assignment to the Minnesota National Guard, he held numerous leadership positions culminating as the Director of Strategic Plans and Policy. In this capacity, he led the MN National Guard Contingency Operations program, which focused on military support to civil authority during national emergencies and national disasters. His team wrote and exercised the plans that provided military resources to civilian authorities and established command authority and relationship development among local, state and tribal emergency response agencies. He oversaw the state partnership program with the country of Croatia, assisting Croatia with the development of various national security programs and policies to include crisis response, critical infrastructure protection and cyber defense training along with traditional military interoperability. He is a graduate of the United States Army War College and the Universities of Minnesota and Wisconsin. He holds an undergraduate degree in earth science education and Masters of Science degrees in strategy and security technologies. He has contributed to academic texts on critical infrastructure protection and written academic papers on energy resiliency. He currently serves the state of Minnesota as an Assistant Commissioner for the Department of Commerce, where he leads a team of 58 consumer service agents and professional investigators. He frequently lectures on cyber security for small businesses and the shared responsibility of the government and the private sector on security and resiliency.

Richard J. Larkin is Former Director of Emergency Management for the City of Saint Paul, Minnesota, USA. He has over 30 years' experience in public safety as an emergency medical technician/paramedic, firefighter, and emergency management practitioner in the 16th largest metropolitan area in the United States. His areas of interest and expertise are emergency management and homeland security program administration, crisis and disaster management, civil protection, critical infrastructure protection and resilience, National Standards and Accreditation of Emergency Management and Business Continuity programs, emergency planning and preparedness, incident management, and emergency response. He is a member of the International Institute for Security Policy and a past Chairperson for an International Emergency Management Standard Development Organization (EMAP).

Sheraz Ali, MBA, is a digital evangelist, serial entrepreneur, guest lecturer, researcher and distinguished international speaker on the subjects of digital transformation, cybersecurity and risk management. He has almost two decades of experience in numerous management and consulting roles in the fields of digital transformation, cybersecurity, risk management, information and technology services, security education training and awareness, business consulting, and financial services. He holds several positions on the advisory boards. He is the Founder and Executive Director of the European Cyber Resilience Research Network (ECRRN), which focuses on building cross-industry bridges to help enhance cyber resiliency of organizations in Europe. He is the Chair of the Board of Directors of The Resiliency Council of Canada (TRCC) committed to helping Canadian industry develop and deliver disruptive intelligence, surveillance and reconnaissance (ISR) technologies to enhance the security and resiliency of the Canadian digital ecosystems. Mr. Ali is also an advisory board member and director of Cybersecurity North America at the International Association of Critical Infrastructure Protection Professionals (IACIPP). Ali holds an Executive Master of Business Administration in Business and Information Technologies from the Nyenrode Business Universiteit in the Netherlands.

INTRODUCTION 1

DOI: 10.4324/9781003273769-1

Chapter 1

Introduction and Methodology

Alessandro Lazari and Robert Mikac

Contents

1.1. Research Background

The subject of CIPR – which in the narrative of this book will be often substituted by the acronym CIP or CIPR – has become important and prominent, especially after all security challenges and changes to the threat scenarios that took place in the first decade of the new millennium, including terrorist attacks (U.S.A. September the 11th, 2001, Spain March the 11th, 2004, U.K. July the 7th, 2005), an increasing number of natural disasters, the global financial crisis in 2008 and the Stuxnet malware which affected the Iranian nuclear facility in Natanz between 2009 and 2010. In addition, with the growing economic and social development, the focus of communities, especially in the context of industrial installations and metropolitan areas, is increasingly directed on maintaining a high level of security in order to preserve functionalities at the local, national and international levels. The field of critical infrastructure, and thus their protection

DOI: 10.4324/9781003273769-2

and resilience, has become an attractive and challenging field for research in the last 15 years due to the multi-faceted aspects of this field: technical, social, legal, and natural sciences. The emergence of new security threats, and the need for a safer society, is on everyone's agenda nowadays and many experts, practitioners, and decision-makers are constantly looking for answers and potential directions to achieve an adequate level of security.

It is a fact that each country in the world depends on critical infrastructure, whether or not they have their own definition of critical infrastructure; established a regulatory national framework; a system for the monitoring of key national risks and processes; a public-private partnership with operators of critical infrastructures and international platform for cooperation. Every country, irrespective of the degree of political, economic, social, technological, normative development, possesses facilities, assets, networks, and systems that are pivotal for the orderly life of modern society, law enforcement and first responders engaged on the issues of public security and safety, up to the defense of the governance of the state.

The importance of various critical infrastructure components grows with the industrial development of global and national communities and depends on the undisturbed functioning of critical infrastructure, both inland and on the territories of other (usually neighboring) countries. The contemporary world has become highly dependent on certain types of critical infrastructure, such as those in energy (e.g. electricity, oil, and gas), telecommunications, transport (e.g. railways, roads, inland waterways, seaborne), banking, and finance sectors. Every disruption in these areas leads to serious malfunctions and difficulties which affect citizens, economies, and states' governance capabilities.

By definition, *"critical infrastructure" represents systems, networks and objects of national importance whose discontinuation of operation or interruption in the delivery of goods or services can have major consequences on national security, health and the lives of people, property and the environment, security and economic stability and continuous functioning of the government.* Critical infrastructure can be defined on multiple levels, but in all definitions, essential elements are the same as in the definition given here. Although it is possible to designate critical infrastructure at many different levels, it is evident from the initial definition that our focus is on infrastructure of national importance, and that will be the primary focus throughout the book. We respect all other approaches and needs, but this research will primarily address issues of critical infrastructure on the state level that are of national importance, while all other reviews will be used secondarily to complete the central interest of the research.

Then, it is important to mention, how in the case of states, there are well-known examples of individual countries that are economically more powerful and territorially larger, and designate critical infrastructure at local, regional and state levels, while smaller and lesser developed countries designate critical infrastructure only at the state level. It is equally important to point out how, critical infrastructure is owned and designated by private corporations and companies, and in the Western world, approximately 85% of national critical infrastructure is privately owned. Therefore, at the beginning of any serious consideration of critical infrastructure issues, public-private partnerships should be perceived and used as an essential model for collaboration and improvement of all processes and functions.

Following the operational definition of critical infrastructure in the preceding paragraph (which will be used throughout the study and analysis in this book), it is important to mention the concept of strengthening the resilience of critical infrastructure. This concept is often forgotten or left out, and mostly the discussion is directed immediately to the area of critical infrastructure protection. However, the concept of strengthening resilience may – but also doesn't have to be – an integral part of the process of critical infrastructure protection. Therefore, it is necessary to emphasize at the outset, what we mean when we refer to strengthening the resilience of critical infrastructure, and, likewise, what we refer to when we use the term "critical infrastructure protection". Strengthening resilience means the use of all prevention measures that must be taken at the design, investment and construction stages – considering all the possible risks and threats. In the exploitation phase, it means the existence of knowledge and ability to manage important processes for the proper functioning of the infrastructure (such as the use of international industrial management standards, own alternative sources of power in the event of a short interruption of external energy supply, diversified key input and output functions and needs, etc.). And finally, cooperation and connection with all relevant external entities and systems that are important for the infrastructure in question to function continuously – to have the necessary inflow resources and raw materials, cooperation with the civil protection system, law enforcement agencies, etc.

On the other side, critical infrastructure protection means:

1) All activities are undertaken primarily by the critical infrastructure owner or manager to prevent internal and external challenges from harming the continuous functioning of the infrastructures.
2) Measures and actions were taken by the state to:
   - establish a normative framework.
   - ensure the development of this area.
   - develop systems for strengthening resilience and protection of key processes and infrastructures in its territory.
   - establish clear competencies and relationships between different stakeholders (system coordinator, competent ministries, security coordinators, regulatory agencies, facilities, networks, and systems that are critical infrastructure).
   - develop public-private partnerships and collaborations within the international community and the organizations to which it belongs or whose membership it seeks.

Here, we clearly see the duality of critical infrastructure protection where both components are extremely important as they ensure the integrity of the process. In this research, we will respect one and the other part, while we will primarily refer, research and describe the part that relates to states – for the simple fact that they are the ones who provide a platform for development and security within the national territory. They also represent the "senior partner" – and often provide regulatory oversight – in the relationship with the owners or managers of critical infrastructures.

An important part of the whole discussion on critical infrastructure is public-private partnership. It has already been mentioned that in the Western world, approximately 85% of national critical infrastructure is privately owned. Therefore, effective strengthening of resilience and critical infrastructure protection, both is a responsibility of the public and private sectors and requires close cooperation between public authorities and the business community. The objective of public-private-sector collaboration in the area of critical infrastructure is to improve investment in infrastructure projects or other activities by providing public risk-sharing services and pooling private sector expertise or additional sources of capital. In this regard, we consider public-private partnerships as an effective way of implementing commitments that ensure the achievement of public policy objectives by linking various forms of both public and private resources and mutual interests.

This book is thematically focused on three key and interconnected segments: first, critical infrastructure as a subject of particular interest for the development of each country and international organization; second, the activities of states which have authority and responsibility within the national territory to regulate (the best they can), the area of strengthening resilience and critical infrastructure protection; third, cooperation between states and the European Union in the field of critical infrastructure by observing it through several different prisms of relations.

1.2. Subject of Research Interest

The challenges in critical infrastructure resilience and protection are numerous. For countries that are just beginning to form the concept of critical infrastructure protection, challenges are manifested in understanding the importance of the aforementioned, building a normative framework, identifying, designating the sectoral and individual critical infrastructure, and establishing the protection measures which are quality and price acceptable, yet appropriate. Then, during the process of designating critical infrastructure, the factors of criticality are extremely important – in terms of national importance and value, such that within critical infrastructure conceptualization there is no critical infrastructure that is not really critical, which can result in the congestion of the process at the very beginning. The next challenge is the relationship between critical infrastructures and their interdependence and criticality, as well as the dependence on the functioning of critical infrastructure on the ground of other states. An additional challenge arises because of the limited institutional capacities and knowledge available for this activity, which greatly complicates a comprehensive approach to critical infrastructure protection and strengthening of resilience. In the context of critical infrastructure protection, internal and external threats pose a special and unique challenge.

Challenges at the European Union level in the process of identifying, defining, and protecting critical infrastructure are multidimensional because it is necessary to harmonize different national policies of Member States and to find and create the identity of the European Union in this area. While some countries such as Germany, Sweden, Great Britain, France, and the Netherlands became advanced in the development of national policies for critical infrastructure protection, the European Union is still searching for its "place and role" in the scope of public policy. The European Commission seeks to promote the importance of the subjected area, to ensure cooperation between the Member States, to accelerate the exchange of knowledge and experience and to guide the Member States in their efforts. At the same time, the European Commission is developing a number of activities in cooperation with neighboring countries (hereinafter in the book under the term "neighboring countries", we will refer to the countries included in EU Enlargement Agenda – Turkey, Bosnia and Herzegovina (BiH), Serbia, Montenegro, North Macedonia, Albania, and Kosova; and EU EaP – Ukraine, Belarus, Moldova, Georgia, Armenia, and Azerbaijan) and transatlantic partners (the United States and Canada) on promoting the importance of strengthening resilience and protecting critical infrastructure. Especially in this part, the European Union is trying to help all neighboring countries to build their own critical infrastructure protection systems. The European Commission organizes meetings and workshops for exchange of information and best practice on critical infrastructure resilience and protection, finances projects involving non-member countries as main beneficiaries, etc.

When observing cooperation at the level of states, it is certain that through multilateral and bilateral efforts, all states are aware of strategic and operational importance of international cooperation on critical infrastructure resilience and protection and cross-border impact, which puts this relation in a high level of focus. For some countries in the Enlargement Agenda and EaP, it could be assumed that the first challenge is the lack of critical infrastructure legislation which provides a definition and models of cooperation – but this should not be an obstacle to the development of the CIP area. We can cite the example of the Republic of Slovenia that has set up and governed the system, which has been in operation for many years, even the *Law on critical infrastructure protection* was eventually adopted at the end of 2017. Everything is ultimately a matter of perspective and a perception of the "state of things" on the national level. Therefore, the European Union and Member states can help neighboring countries a lot, that is, as far as the countries themselves are able, and willing, to use the various cooperation mechanisms that the Union has developed and is willing to share.

The European Union is additionally interested in (for several reasons) strengthening national frameworks to likewise strengthen resilience and the protection of critical infrastructure (both Member States and neighboring countries), since the functioning of the Union itself depends on the functioning of critical infrastructure located in different states, as individual critical infrastructures are (mainly) located in the territory of several different countries. As such, it is holistically necessary to balance resilience and protection in each of the countries sharing the same infrastructure. Also, the Union as a whole is highly dependent on external transport and energy routes, the Union meets more than 50% of its energy needs from imports, through which it provides the necessary resources and raw materials from different parts of the world, and they are extremely important for maintaining the current level of well-being and securing development.

The research interest is to explore the challenges faced by both states and the European Union in the development of the large critical infrastructure arena, and then present the current level of development of this area (at the observed countries and the European Union level), the same as the partnerships that the Union is developing with neighboring countries and transatlantic partners. Finally, we want to provide specific recommendations to everyone involved in this extremely significant process that we all depend on.

1.3. Research Focus

The focus of this research is a multi-level analysis where firstly, the current level of EU achievements in the field of critical infrastructure is presented, then its neighboring policy and transatlantic cooperation, the development of national frameworks of the considered countries, areas of immediate cooperation, but also the shortcomings; this includes suggestions for improving all forms of exchange of knowledge, information and best practices. The research is interdisciplinary and covers several aspects of observation (scientific, legal, and methodological), with a view from the aspect of the public and private sectors and the viability of certain solutions.

This book aims to provide the first comprehensive analysis of the external dimension of the European Programme for Critical Infrastructure Protection. Through a homogeneous methodological framework, the overall aim of the book is to present the current state of development of the neighboring countries of the EU, as well as the transatlantic cooperation of the Union with the USA and Canada. The specific research objectives are directed toward:

1) Providing terms of references to countries that are in the process of establishing/improving a CIPR framework which meets all of today's security preconditions for a safer society

2) Offering insight to institutions and decision-makers on existing policies and ways of improving the European security agenda by taking into consideration the state of play of neighboring countries
3) Gathering different critical infrastructure protection perspectives (development and current state) from the European Union's neighboring countries, plus USA and Canada – whose systems slightly differ and can be a valuable source for the improvement of national and European approaches.

It is a primary goal of this publication to provide an updated picture on how to deal with challenges that CIPR bring. In pursuing such goals, the book will address the following six areas:

1) Shaping of national critical infrastructure protection frameworks, understanding the importance of critical infrastructure service stability and continuity
2) Guidance for building a comprehensive and flexible normative framework
3) Identification of the assets to be protected (criticality) and the establishment of cost-effective permanent and graded measures as part of an Operator security plan
4) Critical infrastructure and dependencies, as well as dependencies from assets placed in neighboring countries
5) Dealing with limited institutional capacities and knowledge available for critical infrastructure protection
6) Strategic and operational importance of international co-operation on critical infrastructure protection.

1.4. Methodological Framework

The central methodological framework of the book is a multi-level analysis where firstly, the current level of EU achievements in the field of critical infrastructure is presented, then its neighboring policy and transatlantic cooperation, development of normative solutions of the analyzed countries, areas of immediate cooperation, but also the shortcomings as well as suggestions for improving all forms of exchange of knowledge, information and best practices.

After the introductory portion explains the structure, background, and focus of the book, the central part of the book presents a set of case studies related to specific countries and how they've developed their policies and processes in the areas of interest. In order for all the case studies to be as homogeneous as possible in their narrative approach, and thus easier to compare, each follows a previously agreed-upon structure and approach:

1) *Introduction* – The introduction will describe the normative framework of critical infrastructure resilience and protection and its link with the European Union's policies and strategies, if any. For "link", it is implied that this section

should clearly highlight whether the national framework has taken any form of inspiration or makes direct reference to the EPCIP policy package (including the *Council Directive 2008/114/EC*) or the Cybersecurity Policy package (including the *Directive (EU) 2016/1148*).

2) *Implementation* – In this part, the national government is represented. This includes the structure, stakeholders, the national plans and the overall system of critical infrastructure resilience and protection, including sections that provide details on how it works and potential gaps.
3) *Public-private partnership* – This section includes description of public-private partnership established in the country and how they work and are regulated, including the activities currently pursued and plans for the future.
4) *Cooperation with European Commission* – The section looks at what cooperation looks like and in what areas of critical infrastructure protection, there is cooperation so far.
5) *Education and training* – This section describes what constitutes formal and informal education of experts who directly and/or indirectly deal with the issues of resilience and protection of critical infrastructure and what types of training and exercises are available in this segment.

The research is interdisciplinary and we cover several aspects of CIP. The assumption and thesis that is to being presented for consideration at the starting point of this research is: The higher the level of cooperation between the European Union – especially the European Commission and the observed countries – the greater the mutual benefit both for the European Union itself and for the countries considered, in strengthening resilience and protecting critical infrastructure. In order to confirm this assumption, we will use several different theories and scientific methods of research. From Neo-functionalism theory, we will use the spillover concept in its functional form, where quality cooperation in one area leads to the extension of cooperation to other areas and promotes political integration (for those countries striving for political integration). From Liberal Intergovernmentalism theory, we will use parts relating to the forms of cooperation between EU institutions and national governments in the field of critical infrastructures, due to their overlapping interests in the field. We will use multi-level governance theory in analyzing the cooperation of national governments with other political actors. Of the scientific methods, we will use the historical method, statistical analysis and synthesis, and the comparative method, through the identification of key indicators.

1.5. The Book Structure

The book is structured into five units and 20 total chapters.

The first unit is the introductory part of the book, which contains an introductory chapter on the methodology and method of making a book. The second chapter describes the current position of European Union Critical Infrastructure External Dimension: from neighboring policy to Transatlantic cooperation. The third chapter is an analysis of the current reach of the Republic of Croatia as the most recent country that has been accepted into the full EU membership and where lessons learned will be presented for neighboring states that are in process of becoming full members of the Union.

The second unit presents an analysis of the development in the area of strengthening resilience and critical infrastructure protection from the side of neighboring countries (EU Enlargement Agenda) where the following countries are involved: Turkey, BiH, Serbia, Montenegro, North Macedonia, Albania, and Kosova.

The third unit is dedicated to the analysis of the activities in countries that are part of EU EaP: Ukraine, Belarus, Moldova, Georgia, Armenia, and Azerbaijan.

The fourth unit contains an analysis of the two countries (USA and Canada) that, together with the European Union, are part of the Transatlantic cooperation in the field of strengthening the resilience and protection of critical infrastructure.

The last, fifth section, a conclusionary chapter, analyzes the procedures of all referenced states from the positions of the European Union and their own positions in the field of critical infrastructure, providing recommendations for continuation of current and future cooperation.

The audience of this book is composed of practitioners in safety and security, business continuity, critical infrastructure operators, experts, and decision-makers in critical infrastructure resilience and protection process who benefits from the overview of best practices and experiences from the countries considered by the book. Also, the audience is composed of all stakeholders, mainly in the public sector (European Commission, national cabinets for critical infrastructure resilience and protection, security agencies), so that can get more acquainted with governance frameworks and their mechanism as well as understand which are the pivotal pillars and requirements currently for ensuring frameworks' performances. Lastly, university students studying safety and security, international relations, crisis management, and political science will find this a useful resource for educational and research purposes, as well as for universities and higher education institutions as part of their curricula.

Chapter 2

The External Dimension of European Union's Critical Infrastructure Protection Programme

From Neighboring Policy to Transatlantic Cooperation

Alessandro Lazari and Robert Mikac

Contents

DOI: 10.4324/9781003273769-3

This chapter describes the European Union's external outreach in the field of CIP, as announced since the initialization of the EPCIP[1]:

> Threats cannot be seen in a purely national context. . . . The interconnected and interdependent nature of today's economy and society means that even a disruption outside of the EU's borders may have a serious impact on the Community and its Member States. Equally true, the disruption or destruction of a critical infrastructure within the EU may have a detrimental effect on the EU's partners. . . . Consequently, enhancing CIP cooperation beyond the EU through such measures as sector specific memoranda of understanding (e.g. on the development of common standards, undertaking joint CIP-related studies, identification of common types of threats and exchanging best-practices on protection measures) and encouraging the raising of CIP standards outside of the EU should therefore be an important element of EPCIP. External cooperation on CIP will primarily focus on the EU's neighbours. Given however the global interconnectedness of certain sectors including ICT and financial markets, a more global approach would be warranted.

The objective of the "external dimension" was crystal clear since the beginning as cooperation beyond the EU's border was deemed necessary to avoid that disruption of critical infrastructures outside of the EU that may have an impact on the EU, and vice versa.

Since the "external dimension" cannot be considered an episodic project but is indeed a long-term objective that nowadays is still being pursued, the objective is to provide an overview of the EPCIP's impact on countries that are a part of the EU Enlargement Agenda, the EU EaP and the Transatlantic cooperation. It is the opinion of the authors, in fact, that the knowledge around the implementation of CIP-related policies and the impact on national frameworks is pivotal for fostering future forms of regional and international cooperation.

Fifteen years after the launch of the EPCIP, the recent history clearly shows how much this program has contributed to the creation of an "EU CIP Community" that has allowed the Member States of the EU, the operators of critical infrastructures and the Academia, among others, to achieve very important objectives that have let the pillars of EU CIP to arise and consolidate. The enablers of these achievements have to be found within the EPCIP which has allowed the Community to work around a "common vocabulary"[2], that have consequently allowed a stronger and sound mutual understanding. These two factors (the community and the common vocabulary) are the sparks that have then allowed the ignition of projects, achievements and harmonization in the areas of governance, cooperation, coordination, solidarity and research. Therefore, it is any envisaged that any impactful cooperation in the domain of the "external dimension" should foresee new cycles of interactions aimed at extending the Community beyond the borders of the EU.

Even though this publication strongly focuses on the EPCIP and therefore on the all-hazard approach on the physical domain of critical infrastructures, the domain of cybersecurity cannot be neglected given its high impact on the security of networks, systems and data that allow critical infrastructures to deliver essential services.

For this reason, two main EU policy packages will be considered:

1. The *European Programme for Critical Infrastructure Protection* (EPCIP) policy package and the *Council Directive 2008/114/EC* of 8 December 2008 "on the identification and designation of European critical infrastructures and the assessment of the need to improve their protection"
2. The EU Cybersecurity policy package and the *Directive (EU) 2016/1148* of the European Parliament and of the Council of 6 July 2016 "concerning measures for a high common level of security of network and information systems across the Union"

It is envisaged that the analysis will most likely include far more commonalities with the Directive from 2008 than the one from 2016 since this last one has entered into force more recently and takes more time to be taken as a reference from neighboring countries.

All the aforementioned policies and directives in the context of critical infrastructures and essential services are part of the EU's efforts to ensure a better environment and achieve a greater level of security in the Member States and the Union as a whole. The structure of this chapter is divided into four sections: 1. introduction, 2. implementation (The normative framework of the European Union in the critical infrastructure protection), 3. cooperation activities with neighboring countries and transatlantic partners, 4. conclusion.

2.1. Introduction

When looking at critical infrastructure and public policy issues related to critical infrastructures in the European Union, it is important to pay attention to several different but complementary dimensions. First, over the last 15 years, the European Union has been progressively developing a strategic and normative framework for this domain. It seeks to harmonize and level the development of the Member States, by supporting those who do not have sufficient governance structure and mechanisms to reach an adequate level of development. It also opens cooperation and knowledge sharing with neighboring countries (EU Enlargement Agenda states and EU EaP states) and transatlantic partners. In addition, the Union has identified four critical infrastructures of the European dimension – Eurocontrol, Galileo, the electricity transmission grid and the gas transmission network as they are recognized to be vital for the Union. These infrastructures have a pan-European footprint, and the European Commission has confirmed its interest in identifying and designating critical entities of "European significance"[3] in the proposal for a Directive of the European Council and the Parliament on the "on the resilience of critical entities", recently published on December 16, 2020.

Although critical infrastructures are important in a number of areas, the vision of strengthening their resilience and protection, and thus the development of the overall dimension of the subject area, has emerged from the security domain. When considering the security strategy documents, in the first European Union Security Strategy, adopted in 2003 under the name *European Security Strategy: A Secure Europe in a Better World*, infrastructures are mentioned in the section on Europe's dependency and vulnerability in the transport, energy, information and other sectors (European Council, 2003). In the following security strategy *Internal Security Strategy for the European Union: Towards a European Security Model* (2010), primarily oriented toward internal security, critical infrastructures are mentioned only once (precisely under the name of critical infrastructure), in the part of their protection in order to ensure a high quality of life in Europe (European Council, 2010). In *Cybersecurity Strategy of the European Union: An Open, Safe and Secure Cyberspace* from 2013, the EU's comprehensive vision and activities in the regulation of cyberspace – as well as the critical infrastructure protection in that dimension (European Commission and High Representative of the European Union for Foreign Affairs and Security Policy, 2013) – was represented. In strategy, *The Global Strategy for the Foreign and Security Policy of the European Union* adopted in 2016 (which replaced the 2003 Strategy), the term and concept of critical infrastructures have been significantly emphasized in the need to strengthen the resilience of infrastructures, investment in their further development and their protection (European External Action Service, 2016). The latest *EU Security Union Strategy* defines four strategic priorities for Union action in the field of security. The latest EU Security Union Strategy defines four strategic priorities for Union action in the field of security. In the first strategic priority, *A future-proof security environment*, CIPR occupies a key place in the consideration and future activities (European Commission, 2020). There is a visible trend in the development of strategic deliberation, in which, over the years, the importance of critical infrastructures has become increasingly emphasized.

In regard to the European Union Critical Infrastructure External Dimension – which includes the EU Enlargement Agenda and the EU EaP and Transatlantic cooperation – it is necessary to provide a short analysis of these policies. The EU Enlargement Agenda is a set of individual policies and bilateral agreements with seven states (Turkey, BiH, Serbia, Montenegro, North Macedonia, Albania, and Kosova) in their process of accession to the Union. With these policies, the EU seeks to assist these countries in their structural reforms, transformation and alignment to the values of the Union. The EU EaP is a joint policy initiative (started in 2009), with the aim to deepen and strengthen relations between the European Union, its Member States and its six Eastern neighbors: Ukraine, Belarus, Moldova,

Georgia, Armenia, and Azerbaijan. The EaP is not aimed at ramping up an EU membership, while the shared objective is to build stronger cooperation in the areas of trade, mobility, economic development and better quality of life. Finally, Transatlantic cooperation is taking place on international priorities with a number of consultations (on several areas of interest) and joint work. The Union and two transatlantic partners (the US and Canada) share many common values and needs, so cooperation between them is necessary. In all three cooperation packages between the Union and the groups of countries mentioned here, there are certain common denominators of collaboration, and one of them is the issue of critical infrastructures – the central interest of this research.

2.2. Implementation

Based on the visions and strategic guidelines set out in the strategies mentioned in the introduction, normative development and implementation of the strategic and operative framework have been initiated and achieved in two parallel and complementary areas, namely: the EPCIP policy package (with *Council Directive 2008/114/EC* as the most significant document) and the cybersecurity policy package (with most relevant, *Directive (EU) 1148/2016*). The first policy package has much more substance because it has been developed over a longer period than the second package, which has seen the light around five years ago and is still being implemented by the Member States of the Union.

2.2.1. European Programme for Critical Infrastructure Protection Policy Package

The path of the European Union in the direction of the protection of critical infrastructure has begun as a response to the terrorist attacks to the US, Madrid and London, that have taken place between 2001 and 2005. For this reason, the EPCIP has always had a strong focus on the protection of critical infrastructures from terrorist attacks. In October 2004, the European Commission adopted the first document in this area: *Communication on Critical Infrastructure Protection in the fight against terrorism*, which presented proposals on what Europe should have done to prevent terrorist attacks on critical infrastructures, to enhance the level of preparedness for emergency situations, to raise their resilience and to develop the ability to respond to attacks (European Commission, 2004). This communication constitutes the initial gateway to CIP for the European Union.

One year later, the Commission published a *Green Paper on a European Programme for Critical Infrastructure Protection*, which provided policy options on how the Commission could establish a critical infrastructure protection program and a Critical Infrastructure Warning Information Network (CIWIN)

(European Commission, 2005). The discussions that were conducted after the adoption of the *Green Paper* highlighted the added value of setting up the Union's strategic framework for critical infrastructure protection. Also, the key directions of the development of this area are highlighted, such as the need to improve capabilities for critical infrastructure protection in Europe and the need to help alleviate weaknesses related to critical infrastructure. Furthermore, the importance of key principles of subsidiarity, proportionality and complementarity have been highlighted as well as the dialogue between stakeholders in the system of strengthening the resilience and critical infrastructure protection (Council of the European Union, 2008).

The next input came from the Justice and Home Affairs Council, which in December 2005 called upon the Commission to make a proposal for a *European Programme for Critical Infrastructure Protection*. Therefore, in 2006, the European Commission proposed a *European Programme for Critical Infrastructure Protection*, which embraces an all-hazards approach even if terrorism remains the primary focus (European Commission, 2006).

This document was updated in 2013 with a proposal for a new approach: *Commission Staff Working Document on a new approach to the European Programme for Critical Infrastructure Protection: Making European Critical Infrastructures more secure*, where the emphasis is in the areas of prevention, preparedness and response, as well as greater collaboration with various partners and projects (European Commission, 2013).

In April 2007, the Council of the European Union published some conclusions regarding the program, stating that the ultimate responsibility for protecting critical infrastructures lies primarily on the Member States, within their national borders. In addition to this, the Council called for an effort from the Commission to develop a procedure for the identification and designation of ECIs and the assessment of the need to improve their protection. This work was seen as a priority, as it was recognized that there were a number of critical infrastructures in the Union in which disruption or destruction could have led to significant cross-border effects. With this suggestion, the Council recognized that an integrated solution adopted on the entire Union was necessary since bilateral cooperation programs between the Member States in the area of critical infrastructure protection were not enough for properly dealing with CIP at the Union level (Council of the European Union, 2008).

After that, the Council of the European Union, taking into account the proposal of the Commission, immediately developed a key proposal for the area of critical infrastructures in the European Union, which led to the promulgation of the *Council Directive 2008/114/EC of 8 December 2008 on the identification and designation of European critical infrastructures and the assessment of the need to improve their protection* (further *Council Directive 2008/114/EC* or *Directive 2008/114/EC*), which, as much as the EPCIP, is no longer primarily focused on the threat of terrorism, but seeks to establish a comprehensive process of critical infrastructure protection at the level of both the Member States and the Union as a whole (Council of the European Union, 2008).

Council Directive 2008/114/EC

As mentioned in the introduction of this chapter, all along the 15 years of work around CIP, the Member States have initially achieved the establishment of a community and together with it, the definition of a common vocabulary, which has enabled greater awareness and common understanding around key issues of European CIP. The second objective has been achieved thanks to the adoption of a common definition, which has been included in the text of the *Directive 114/08/EC*, which states that critical infrastructure means

> an asset, system or part thereof located in Member States which is essential for the maintenance of vital societal functions, health, safety, security, economic or social well-being of people, and the disruption or destruction of which would have a significant impact in a Member State as a result of the failure to maintain those functions.

At the same time, the Directive defines ECI as

> critical infrastructure located in Member States the disruption or destruction of which would have a significant impact on at least two Member States. The significance of the impact shall be assessed in terms of cross-cutting criteria. This includes effects resulting from cross-sector dependencies on other types of infrastructure.
>
> (Council of the European Union, 2008)

These two definitions are still in use as of today, since *Directive 114/08/EC* is still in force. The plans for its substitution have been unveiled in December 2020, when the European Commission has published a new proposal for a Directive based on the "resilience of critical entities" which is being discussed in both the Parliament and the Council at the time this publication is being released.

The Council Directive 2008/114/EC represents the central point of EU policy development for the Member States, accession countries and candidate countries for EU membership and provides the logic behind establishing business processes and the basis for a number of other activities (such as EU-funded projects, development of cooperation between states and critical infrastructure operators, the establishment of public-private partnerships, development of professional curricula, foundation of centers and summer schools with a special interest in critical infrastructures). For that reason, an in-depth analysis will be provided in the following text.

In the introductory provisions of *Council Directive 2008/114/EC*, the essential guidelines are presented to set the scene around the expected core pillars. It was emphasized that the first step in the multiphase approach is aimed at the identification and designation of ECIs and the assessment of the need to improve their protection. Even though in the initial discussions and in the *Green Paper* the project was to include at least eight sectors, to define the categories of critical infrastructures of interest, in the final text of the Directive the focus is primarily set on the sectors of energy and transport. As anticipated earlier, what is especially important, is that the Member States and the owners or operators have the primary and ultimate responsibility for the critical infrastructure protection in their respective nations and in Europe.

Another important aspect of *Council Directive 2008/114/EC* is that it has become a common platform for the cooperation of all relevant stakeholders of the critical infrastructure protection domain at the Union level. Prior to its adoption, the obligation of official cooperation among various stakeholders (public bodies and operators), as well as the forum for achieving this cooperation, did not exist. After 15 years, it can be said that this platform is among the most important achievement of the combined effort of the EPCIP and the Directive since it has led to the establishment of a collaboration, coordination and cooperation platform that has no other similar in the current landscape. The same platform, thanks to the number of consultations, discussions and projects, has also created the basis for the promulgation of the so-called NIS Directive which has profoundly benefited from the activities carried on within the CIP context.

The central part of *Council Directive 2008/114/EC* is to be considered the procedure for identification and designation of ECIs. The identification procedure was adopted in article 3 and the accompanying attachments. It consists of several steps involving a number of definitions and cross-cutting (e.g. casualties, economic losses, loss of public confidence) and sectoral criteria. Among the steps to be covered in order to properly identify and designate ECIs, it is worth mentioning the assessment of the cross-border impacts which aim to determine whether the disruption or destruction of a certain infrastructure is mutually significant for two or more Member States. The concept of "trans-boundary externalities" is probably the most important lesson of the Union CIP's legacy that has led to the establishment of a mechanism that should be duly considered by every country. It is the opinion of the authors, in fact, that every country should always assess incoming and outgoing trans-boundary externalities as it is a fact that no country is independent in areas like – for example – energy and food supply chains, transports and telecommunications.

Given the complexity of the subject, the European Commission offered assistance to the Member States in identifying potential critical European infrastructures in case of lack of administrative and professional capacity, lack of established procedures or uncertainty in the interpretation of certain criteria. The Commission, in fact, has since 2009 played an active role in this area by organizing a number of meetings, activities, projects as well as experimental research to foster the process of harmonization and improve the acquaintance of the stakeholder with the peculiarity of this rather complex domain. Among the activities to be noted, the plenary meetings of the Point of Contact (PoC) of the Member States were organized biannually by DG Home, in which the Member States have shared their views in the implementation of the Directive, including lessons learned, best practices and national approaches. In this specific area, the more industrialized and advanced Member States have played a very important role by showcasing their already established frameworks and the means they had adopted to include the requirements of the Directive. As part of the "external dimension" pillar of the EPCIP, some of these meetings have also included an enlarged audience, with the inclusion of Points of Contacts coming from countries of the Second Enlargement Agenda and EaP.

Following the successful negotiations between the Member States, regarding the designation of an ECI (as described in article 4 of the Directive), the next step is to inform the owner or operator of the critical infrastructure that his infrastructure has been designated as an ECI. The Member State on whose territory this ECI is located is responsible for informing the owner or operator and is also obliged on annual basis to inform the Commission of the number of designated ECIs per sector and of the number of Member States dependent on each designated ECI. The information on the designated infrastructure is classified and their identity is known only between the Member States that shares mentioned and/or are in any way dependent on it.

Another pivotal step foreseen by the Directive is the one related to the preparation of operator security plans or equivalent documents that have to include the identification of important assets, the risk assessment and the selection and prioritization of countermeasures, and procedures for the protection of those assets. In order to avoid unnecessary work and duplication of documents, each Member State should first determine whether owners or operators of the designated ECI have already established OSP or other equivalent documents. Where such plans exist, it is necessary to analyze them and see if they need to be upgraded; and where they do not exist, each Member State should take the necessary measures to ensure their preparation.

The final important provision is the designation of a Security Liaison Officer (SLO). This measure has been foreseen to guarantee that each owner or operator has appointed a security coordinator who is the qualified expert in charge of the information exchange between national and European bodies, that, at the same time, has an important vertical role within the company, to advocate the adoption of the necessary protection measures.

Council Directive 2008/114/EC was originally used to guide Member States in their mutual cooperation and provide an example of how they can directly establish and organize the national framework for identification and designation of critical infrastructures and indirectly for their protection. The national responses to the *Directive 2008/114/EC* have set a number of practical solutions that serve to help the states design and structure their internal processes related to national critical infrastructures protection. A clear example of the thorough embracement of the European values and procedures on CIP can be found in the legislative framework of the Republic of Croatia. In this case, the legislator decided to fully follow the narration and content of *Council Directive 2008/114/EC* in the development of the *Critical Infrastructure Law*. Many other Member States have embraced a similar approach by establishing at the same time a procedure for the identification of both national and ECI. Such an approach, in the cases of countries which did not have any framework of law on CIP, has led to the establishment of a very focused framework which has shown very limited bureaucracy.[4]

2.2.2. Cybersecurity Policy Package

Since critical infrastructures are connected and increasingly dependent on the Internet and services delivered in cyberspace, the Union has taken steps to regulate also this domain. In 2013, the European Commission, together with the High Representative of the European Union for Foreign Affairs and Security Policy, has published the *Cybersecurity Strategy of the European Union: An Open, Safe and Secure Cyberspace,* which represented the EU's comprehensive vision on how to best support Member States and other stakeholders in preventing and responding to cyber disruptions and attacks (European Commission and High Representative of the European Union for Foreign Affairs and Security Policy, 2013). The vision was

> to foster European values of freedom and democracy and to ensure that the digital economy can safely grow. Specific actions aimed at enhancing cyber resilience of information systems, reducing cybercrime and strengthening EU international cybersecurity and cyber defence policy.

The strategy articulates the EU's vision of cybersecurity through five priorities:

1. Achieving Cyber Resilience
2. Drastically reducing cybercrime
3. Developing cyber-defence policy and capabilities related to the Common Security and Defence Policy (CSDP)
4. Developing the industrial and technological resources for cybersecurity
5. Establishing a coherent international cyberspace policy for the European Union and promoting core EU values.

According to the strategy, the roadmap is implemented through the following instruments: legislative instrument, non-legislative instruments and funding activities (European Commission, 2017: 2–3). The strategy is an essential basis for further joint activities in the regulation of cyberspace as well as critical infrastructure protection in that dimension because "securing network and information systems in the European Union is essential to keep the online economy running and to ensure prosperity" (European Commission, 2019).

Based on the aforementioned strategy, the *Directive (EU) 2016/1148 of the European Parliament and of the Council concerning measures for a high common level of security of network and information systems across the Union* (further *NIS Directive*) was adopted. It was adopted on July 6, 2016, with the obligation to be implemented into national legislation of all Member States until May 9, 2018 (European Parliament and of the Council, 2016). The *NIS Directive* represents the main legislation on cybersecurity in the Union as the time this publication is being released. Plans for its amendment have already been published in December 2020 by the European Commission with the unveiling of a new proposal for the NIS 2.0. To this end, the Commission proposal

> expands the scope of the current NIS Directive by adding new sectors based on their criticality for the economy and society, and by introducing a clear size cap – meaning that all medium and large companies in selected sectors will be included in the scope. The proposal also eliminates the distinction between operators of essential services and digital service providers. Entities would be classified based on their importance, and divided respectively in essential and important categories with the consequence of being subjected to different supervisory regimes.[5]

NIS Directive

NIS and services play a vital role in society. That is why the *NIS Directive* was adopted to connect the key areas, actors and processes, in order to increase the level of protection and the introduction of minimum common standards in the area of cybersecurity.

The *NIS Directive* covers two groups of actors: operators of essential services and digital service providers. Operators of essential services are considered those who provide key services to society or the national economy in the following seven sectors: energy, transport, banking, financial market, health, drinking water supply and distribution, and digital infrastructure. Digital service providers are considered to be of general importance when it comes to cybersecurity and include providers in the following three sectors: marketplaces, cloud computing services, and online search engines.

The main objective of the *NIS Directive* is to provide a common level of security of NIS in all Member States, mindful that their malfunctions (due to security incidents) could have significant consequences on the national economy or society as a whole. *NIS Directive* introduces regulatory elements which are enabling permanent monitoring of the condition of automation and digitization of defined sectors. In addition, it stipulates the obligation to implement technical and organizational measures for risk management and also measures to prevent and minimize the effect of the incident on the security of NIS. This Directive also introduces an obligation for operators to notify about incidents that happened in or on systems that may have a significant effect on the continuity of providing specific service.

The *NIS Directive* prescribes an obligation for Member States to acquire a high level of protection for cybersecurity (by introducing measures to achieve it), in the following sectors:

- energy (electricity, oil, gas)
- transport (air, rail, water, road)
- banking
- financial market infrastructures
- health sector
- drinking water supply and distribution
- digital infrastructure (internet traffic exchange, domain name services and national top-level domain control).

For the purpose of implementing the objectives of the *NIS Directive*, all Member States were also obligated to adopt a national strategy for the security of NIS.

The *NIS Directive* is part of a broad EU initiative aimed at improving the security of systems and networks on which operators of essential service rely to. Thereby, the *NIS Directive* creates an appropriate framework for the prevention and protection of society against cyber threats by establishing a common approach of all Member States, as they individually ensure harmonized vertical sectoral approaches in terms of the *NIS Directive*, while the new EU Personal Data Protection Regulation (GDPR) provides a similar horizontal approach throughout all segments of society as a whole.

Observing the *NIS Directive* in relation to *Council Directive 2008/114/EC*, it is necessary to highlight several important issues. We can say that the *NIS Directive* has been developed from the need to complement the normative framework, because of the lack of adequate critical infrastructure protection and operations in the information and communication technology (ICT) sectors. *Council Directive 2008/114/EC* not only focuses primarily on energy and transport sectors but also emphasizes the need that other significant sectors, such as ICT sectors, to be considered. While operators of critical infrastructures and operators of essential services do not necessarily coincide, there is also a great likelihood that they will overlap in many cases. *Directive 2008/114/EC* is more focused on assets, while the *NIS Directive* is more focused on services. The main objective of the *Directive 2008/114/EC* is limited to enhancing the security of specific critical infrastructures that are important at the EU level, while on the other hand the *NIS Directive's* main objective is to enhance the EU–Member States security via Member States security and EU cooperation.

2.3. Cooperation with Member States, Neighboring Countries and Transatlantic Partners

Throughout the last 15 years, the European Union, especially European Commission, has put in place an unprecedented effort aimed at the establishment and development of different cooperation platforms between Member States, neighboring countries, and transatlantic partners. The Commission is also cooperating, and endeavoring to cooperate, with owners or operators of critical infrastructures, international organizations, and interested experts. This research is focused on the collaboration that takes place between the European Commission and the Member States, neighboring countries, and transatlantic partners. In this domain, as mentioned at the beginning of this chapter, we recognize a successful model: the meetings of the PoC of the Member States, the meetings with neighboring countries and transatlantic partners, the great cooperation platform established through ERNCIP and the EU-funded projects (e.g. CIPS, FP7, H2020).

As anticipated, when it comes to direct collaboration between the European Commission and the Member States, one of the most concrete activities is the gathering of national contact points for CIP in the plenary meeting taking place twice per year.

This model of cooperation has proven to be very successful, so the Commission has decided to apply it to cooperation with neighboring countries, transatlantic partners and also on the Cyber pillar constituted by the *NIS Directive* (where similar meetings regularly take place). EU–neighboring countries meetings are part of the external domain of the *European programme for critical infrastructure protection* (EPCIP), aimed to deepen CIP-related forms of cooperation. To foster strategic partnerships beyond Europe, EU–US and EU–Canada expert meetings have been held yearly. These meetings addressed mainly the need to strengthen cooperation by sharing knowledge, best practices and information on CIP, including the development of a global infrastructure security toolkit (European Commission, 2013).

In addition to this formal network, the Commission strongly encourages the Member States to take part in the European Reference Network for Critical Infrastructure Protection – ERNCIP), which is a network of experts in key areas of CIP like CBRNe, physical security and hybrid threats. The network aims to provide a framework within experimental facilities and laboratories to share knowledge and expertise to harmonize protocols, standards and procedures across Europe. The goal is for such collaboration to lead to better protection of critical infrastructure from all types of threats and hazards.

Finally, the European Commission is co-financing a number of projects in the field of critical infrastructure protection. The main purpose of project financing is to improve knowledge, provide a better understanding of the functioning of critical infrastructures (at all levels) and provide recommendations for the development of public policies. The option of using EU funds for CI-related projects is strongly recommended to all countries that are at the beginning of the establishment of their national CIP frameworks or want to improve the existing ones.

Chapter Conclusion

Since the first policy on countering terrorism launched back in 2004, the European Commission has constantly amended and upgraded the European agenda on security. This approach has allowed the deployment of new packages aimed at providing an adequate pace to the European roadmap so that most of the threats are addressed. In the field of critical infrastructures, the EPCIP is a milestone since it has created awareness throughout the Union on the need to establish national frameworks on CIP and work together with operators of CIs in strengthening the vital services which are deemed necessary for the security of citizens and the stability and growth of the Union. In the coming chapters, the impact of the EPCIP and the other policy considered by this publication will be analyzed in detail by taking into account perspectives from EU's neighboring countries and partners from the eastern and transatlantic cooperation. As already mentioned in the introduction of this chapter, since the EPCIP and the *Council Directive 2008/114/EC* are about to be amended, the editors' intent is to provide the policymakers with a critical mass of information that could pave the way toward a better policy and international cooperation by offering a map of the current status of neighbors and transatlantic allies and partners.

As a last note, it is the opinion of the editors that areas – which are not the focus of this publication but constitute milestones from which other important lessons could be drawn – are the ones related to Seveso, airport and maritime security and the Civil Protection mechanism. The security agenda of the Union, in fact, goes way beyond the EPCIP and can also rely on very mature and established policies and legislation.

Notes

1. See the chapter 7 of the Communication from the European Commission "on a European Programme for Critical Infrastructure Protection" COM (2006) 768 final from the 12.12.2006.
2. "Meant as the attitude to a 'better mutual understanding' and convergence in the interpretation of technical word that are slowly becoming part of the 'European CI' vocabulary" as in A Lazari, European Critical Infrastructures, Springer, 2014.

3. At the time this publication is being drafted, the version of article 14, which introduces the concept of Critical Entities of European Relevance, states the following: "An entity shall be considered a critical entity of particular European significance when it has been identified as a critical entity and it provides essential services to or in more than one third of Member States."
4. Meant as projects or activities which do not lead to direct impacts on the improvement of the protection of a critical infrastructure because of the presence of too many procedures.
5. A full description of the new proposal and its alignment with the EU's Cyber Strategy is available here: https://digital-strategy.ec.europa.eu/en/library/proposal-directive-measures-high-common-level-cybersecurity-across-union.

Chapter 3

Croatia

Terms of Reference for Neighboring Countries

Ivana Cesarec

Contents

This chapter provides an insight into the efforts and achievements of the development of this security field in the Republic of Croatia. From establishing a normative framework as the basis for the identification and designation of critical infrastructure (CI), through the activities carried out to create an optimal critical infrastructure protection (CIP) system – it comprehends open issues and challenges facing each country, especially those countries that have not considered an establishment of such system so far. Croatia's experience in developing and aligning its national framework with European strategies and policies can be a good reference for Neighboring countries and EaP countries, who are not yet EU members and just beginning the process of forming the CIP system.

DOI: 10.4324/9781003273769-4

3.1. Introduction

The Republic of Croatia is the latest country to accede to the European Union in 2013, and through the process of becoming a Member State had an obligation to normatively arrange and regulate issues of identification, designation and protection of ECI. These obligations emerge from *Council Directive 2008/114/EC on the Identification and Designation of European Critical Infrastructures and the Assessment of the Need to Improve their Protection* (*Council Directive 2008/114/EC*). The Republic of Croatia has implemented *Council Directive 2008/114/EC* into national law by drafting and adopting the *Critical Infrastructure Law* (adopted in 2013), which provides regulation for national and European CI (as a comprehensive solution in response to requirements from the European Commission).

Before this, the Republic of Croatia has undertaken strategic and normative implementation activities that have signified the importance of critical infrastructure and the need for its adequate protection. Some of those strategic documents are the *National Strategy for the Prevention and Countering of Terrorism* (2008), where the critical infrastructures concept was perceived from the aspect of protection against terrorist threats and the strategy points out that the Republic of Croatia needs to build national capabilities to protect critical infrastructure.[1] The 2010 *Protection and Rescue Plan for the Republic of Croatia* prescribes obligations (protection and rescue measures) of the participants of the civil protection system through the determinants of protecting vulnerable (endangered) critical infrastructure facilities (in case of flooding) and restoring critical infrastructures' facility functions (in case of earthquakes); *Risk Assessment for the Republic of Croatia from Natural and Technical-Technological Disasters and Major Accidents* (2013) puts critical infrastructure in a wider range of protection from natural and anthropogenic threat sources; and *The National Strategy and Action Plan for the Non-proliferation of Weapons of Mass Destruction* (2013) mentions the protection of critical infrastructure and the population from the crisis caused by the mass destruction as one of the specific objectives. These documents showed the necessity for forming the CIP framework as a normative framing of concepts related to critical infrastructure.

Before the 2013 *Critical Infrastructure Law*, the normative implementation issue of the protection of critical infrastructures (or objects) significant for the state and society was covered by several different legal solutions. These are the *Police Law*, which in every state regulates the police obligation to protect particularly important objects, and the set of acts governing the activities of an intelligence system that has the task of preventive action to ensure that critical infrastructure and particularly significant facilities are not endangered by individuals or groups who have undesirable intentions. But here it is necessary to specifically mention two normative solutions and processes that are particularly significant for the focus of the research topic. First, Croatia has long established the process of identifying and protecting objects especially important for the defense of the country.[2] In that process, (which was launched in 1999), by the adoption of the *Rules on Criteria for Designating and Protecting Objects Particularly Significant for the Defense of the Country* under the *Defense Law*, there is defined methodology for threat assessment and protection plan of military and other objects which are conclusively designated. The coordinator of that system is the Ministry of Defense that manages a well-developed, functional system. For a large number of objects, it can be assumed that it will be overlapped with potential critical infrastructure objects because of the great similarity (in some cases, the equivalence) of the stipulated provisions with those of the *Critical Infrastructure Law*. Two parallel processes emerge for objects that are equally important to be recognized as objects of special interest to the defense of the country and as the CI objects, with two different system coordinators (Ministry of Defense since 1999 and previously competent NPRD since 2013 when *Critical Infrastructure Law* was adopted). The second one is *Private Protection Law*, where critical infrastructure is defined as follows: "Critical infrastructures are activities, networks, services and goods of material and information technology whose malfunction or destruction would significantly affect the health and safety of citizens or the effective functioning of the State government" (Croatian Parliament, 2010: article 2). The state stipulates that critical infrastructure should be protected but the owner/manager decides how. Since private security companies have significant capacities to protect the property (human and technical solutions), they are engaged to ensure a high level of security, primarily in the preventive segment, and in accordance with the *Private Protection Law*. The public-private partnership is addressed in particular in the "security privatization" segment. It is conditioned on a clear definition of the roles and responsibilities, based on the regulation and the specific contractual obligations defined by the managers of critical infrastructures.

As we can observe, before joining the European Union, Croatia did pay attention to critical infrastructure protection in strategic documents and certain normative acts in the field of defense and security, but it did not form a complete framework of laws and other regulations to begin the process of CIP system development. In this period, the definition of critical infrastructures was also adopted, and all valuable and exceptionally important objects, networks and systems were recognized and protected (by some other processes and legal obligations), although they were not called "critical infrastructure" in the spirit and sense of *Council Directive 2008/114/EC*. Therefore, in order to "round up" all national efforts and commitments to the European Union, the process of reconceptualization and the setting up of a new system that will concern the area of critical infrastructures under the Union guidelines has been initiated. The process began with *Critical Infrastructure Law* adoption, and it is important to point out that in addition to opening a new process, introducing new critical infrastructure definitions and models of cooperation, the process also brings an important emphasis on the field of resilience strengthening and protecting national critical infrastructures.

Under this Law, national critical infrastructure is defined as

> the systems, networks and objects of national importance whose disruption in operation or interruption in the delivery of the goods can have serious consequences for national security, health and lives of people, property or environment, security and economic stability and continuous functioning of the Government.
>
> (Croatian Parliament, 2013: article 3)

The *Critical Infrastructure Law* has determined the rights, authorities and obligations of the Government of the Republic of Croatia and state administrative bodies, as well as owners/managers of CI – necessary for identifying, designating and protecting national CI and securing their uninterrupted functionality. Also, the Law determines CI management and risk analysis, regulates owners'/managers' Security plans, the role of CI security coordinator, procedures for handling sensitive information, data classification processes, and the supervision of implementing the Law's provisions.

Today, in the normative CI framework of the Republic of Croatia, with the *Critical Infrastructure Law*, there are also two equally important subordinate documents: *Decision on Designation of the Sectors from Which the Central State Administrative Bodies Identify National Critical Infrastructure and Lists of the Order of the Sectors of Critical Infrastructures* (Decision on Designation) and the *Rules on the Methodology for Drafting Business Risk Analysis of Critical Infrastructure* (Rules on the Methodology). In the Decision on Designation, a total of 11 sectors have been determined from which ministries (the central administrative bodies) can identify the national CI. Those sectors are energy, communications and IT, transport, public health, water management, food, finances, production, storage and transport of hazardous materials, public sector, national monuments and valuables, and science and education. For these sectors, there are nine competent ministries,[3] along with the Ministry of Interior (MoI), Civil Protection Directorate (which is a legal successor of state administrative body National Protection and Rescue Directorate – NPRD).[4] Since NPRD was (very recently) the coordinator of the CIP system (and the national contact point for cooperation with other countries and the European Commission), in this overview, it will be mentioned in that role. The second document, the Rules on the Methodology, determines the guidelines, criteria and measures for CI identification and risk analysis management. In order to improve and harmonize with international standards (specifically *ISO 31000: 2009 Risk Management – Principles and Guidelines*), a new *Rules on the methodology for drafting business risk analysis of critical infrastructure* was adopted and implemented in 2016, which replaced the 2013 Rules on the Methodology. This shows that the Republic of Croatia is constantly working on the improvement of the CIP normative framework.

The existing normative structure, with the implementation of necessary actions, as well as cooperation and coordination, has proved to be challenging. Considering the implementation of *Council Directive 2008/114/EC* into the Croatian legislation, the possibility of identifying and designating critical infrastructure in even 11 sectors is a large task, bearing in mind that *Directive 2008/114/EC* obliged all the Member States to take into consideration two sectors: energy and transport, while the inclusion of other sectors was optional for Member States. This is something that needs to be taken into consideration for all countries that are just starting with this process. In the next subchapter, the implementation (and linked challenges) of the regulatory framework established with the relation with the European Union's security policy, will be identified and shown.

3.2. Implementation

The Republic of Croatia, as mentioned earlier, had established a normative framework in accordance with the *Council Directive 2008/114/EC*, had set up the architecture of the CIP system, organized training for security coordinators in the competent central state administration bodies to support stakeholders in the implementation of commitments, held meetings with Republic of Slovenia and Hungary (regarding the identification of ECI) and implemented the RECIPE 2015 project (funded by the European Union to support further development of initiated system-building activities). Identification of critical infrastructure (as a starting point for the development of a critical infrastructure protection system) had its challenges since other (linked) processes such as defining cross-sectoral criteria were long only in the discussion phase. With the adoption of decision on cross-sectoral criteria in 2016, the process became more operational.

The importance of raising CIP system significance on a national level actualized in recent years by following the implementation of the *European Programme for Critical Infrastructure Protection* policy package and *EU Cybersecurity* policy package. Regarding the first one, in 2017, two important national security documents were adopted, which introduced critical infrastructure on the list of national priorities – *Republic of Croatia National Security Strategy* and the *Homeland Security Law*. The Strategy, in particular, through the nine strategic goals for the Republic of Croatia (which represent a concrete realization of the achievement of the national security policy) underlines the strategic objective of "achieving the highest level of security and protection of the population and infrastructure", derived from one of the four fundamental national interests: "the safety of the population, territorial integrity and sovereignty of the Republic of Croatia" (Croatian Parliament, 2017a). The aim of the *Homeland Security Law* is to establish an effective Homeland Security System in response to contemporary threats and risks by building a secure environment and protecting and promoting the national interests of the Republic of Croatia, taking into account, protection of the people, economy, infrastructure and natural resources of the state territory as a prerequisite for the overall prosperity of the Republic of Croatia and the well-being of its citizens (Croatian Parliament, 2017b).

In response to the EU Cybersecurity policy package, a fully fledged *National Cyber Security Strategy* (2015) was adopted in line with the EU's principles and objectives. The Strategy has also provided general orientation for priorities (Government of the Republic of Croatia, 2015b). After that, there was a transposition of the *NIS Directive* through the 2018 *Law on the Cyber Security of the Key Service Operators and Digital Services Providers*. The aim of the Law is to ensure a high level of cyber security in providing services, important for carrying out the key social and economic activities. The Law has prescribed criteria and thresholds for assessing the importance of the negative impact of an incident and those were used as the basis for making cross-sectoral criteria for the needs of national CI identification as well (Croatian Parliament, 2018).

Such strategic direction in the Republic of Croatia and the setting of national priorities is ensuring the easier establishment of an adequate platform for the development of a critical infrastructure protection system. Then, the same pattern and logic of development of this area are visible in accordance with the guidelines promoted by the European Union (which is logical because Croatian experts participate in the work of different bodies of the Union in these areas). Similarly, there has been a higher level of cooperation between the different state bodies in charge of drafting normative solutions and coordinating different processes. Additionally, the CIP system emphasis is getting clearer in 2019, with the constitution of the Civil Protection Directorate of MoI, which formed the special unit named Critical Infrastructure and Cultural Heritage Department, which is primarily dealing with the tasks and activities of CIP. This strengthens the position of system coordinator which in the previous setup (as state administrative body) did not have the expected institutional power. The involvement of the ministries in charge of the identified CI sectors is also renewed. Their contribution is fundamental for the preparation of an initial list of national critical infrastructures to be later validated by the Government. Also, the process of revising the *Critical Infrastructure Law* has begun, and there will be a special focus on the inclusion of all relevant CIP system stakeholders in order to participate in the formation of the most adequate system.

In relation to stakeholder cooperation, a key component is a public-private partnership. The establishment of quality cooperation will be discussed in the following subchapter.

3.3. Public-Private Partnership

The private sector, which is most often the owner and the critical infrastructure manager (for example, Croatian Telecom[5] in the communication and IT sector), has the responsibility to protect the infrastructure that is important to the functionality of the entire society, and this cannot be done efficiently and without greater cost, if there is no cooperation with public institutions (regardless of the form of ownership – effective protection of critical infrastructure is a task and responsibility of both the public and private sectors and requires narrow cooperation between public authorities and the business community). Such relationship creates a number of open issues, like developing common procedures, defining roles and responsibilities (regulated by legislative), exchanging sensitive data related to the building of trust (and the protection of sensitive data, since the private sector, by revealing information/data that is potentially trade secrets, may harm its competitiveness), exchanging knowledge and experiences. That is why, in the Republic of Croatia, it is necessary (in addition to existing public-private partnership models) to work on further strengthening and greater public-private cooperation. This is extremely important because it is not only a matter of obligations that the state prescribes for the private sector, yet if we take just an example of the aforementioned case of Croatian Telecom, which is part of the Deutsche Telecom AG Corporation – Croatian Telecom applies all the procedures, standards and knowledge accepted at the level of the entire Deutsche Telecom AG Corporation and they are most certainly on a higher level, compared to a country such as the Republic of Croatia, which has just begun to establish its system of strengthening resilience and protecting critical infrastructures. So, through an open partnership with the private sector, the public sector can learn a lot, set up a better system and manage it with greater quality.

The Croatian normative model of public-private partnership has already been mentioned in part of referring to certain normative solutions and, in particular, in accordance with the obligations set out in the *Private Protection Law*, which partly relates to the technical and physical protection of infrastructures as the interests of our research. In the Republic of Croatia, since 2012, exists the *Public-Private Partnership Law* (with 2014 and 2018 amendments) which also partially regulates the public-private partnership area that we analyze. However, this Law is currently regulatorily limited to investments for the construction and maintenance of objects (facilities) and is not adapted to the overall needs of the critical infrastructure field. Therefore, it is necessary to change and establish an acceptable public-private partnership model of cooperation in different CI areas not just in construction and maintenance, with clearly defined mutual rights and obligations in order to be easily achievable and of interest to both parties. The normative framework should not only allow the regulation but also leave sufficient space for new solutions and ways of implementing legal provisions. Changes/amendments to the *Critical Infrastructure Law* should also implement the concept of public-private partnership, which is currently missing. That is something that all currently active stakeholders in Croatia are aware of, and it has been part of many joint discussions.

The exchange of stakeholders' opinions as well as knowledge and experiences is facilitated by many national (and international) conferences organized in the past years in the Republic of Croatia not only by the national coordinator of the CIP system but also by other security-related institutions, such as Croatian Institute for Urban Security, Journal Zaštita and RACVIAC – Centre for Security Cooperation (for example: "Challenges in industry and infrastructure protection" in 2019, "Public and private security companies in critical infrastructure protection" in 2018, "Public-private partnership in critical infrastructure protection" in 2017, "International conference on CIP" in 2016 and so on).

In the Republic of Croatia, a good example of using private protection and public-private partnerships on facilities important for the functioning of the state are the tasks of passengers and luggage control at the biggest Croatian international airport, Franjo Tuđman, which is done by private security companies (and which was previously the task of the employees of the MoI). The second good example is JANAF[6] – an external company (private security company), which provides physical and technical protection to JANAF's objects. The security company works together with JANAF employees in control centers at the Terminal Sisak and Terminal Omišalj, and in the operational and communication center at Zagreb, where everyone is responsible for their part of the work and have a common situational image, giving them a faster possibility of reaction.

One of the greater challenges represents companies that are majority-owned by the state and are primarily focused on meeting narrow political and economic goals and therefore do not consider the business continuity of critical infrastructures to the expected extent. On the other hand, some private companies, focused primarily on economic profit, have developed mechanisms to maintain business continuity, but do not invest enough in critical infrastructure protection measures, as this is not their strategic goal (and they also require additional investment). For this reason, it is imperative to raise awareness of the need to protect their infrastructures, knowing what the consequences of CI disruption are.

3.4. Cooperation with European Commission

The Republic of Croatia is following the processes and developments of the European Union and individual Member States for achieving its own national CI security and resilience – especially, having in focus the *Council Directive 2008/114/EC*, which has become a platform for the cooperation of all relevant stakeholders of the critical infrastructure protection system at Union level. Before the directive, there was no obligation to formally cooperate with different stakeholders, nor a forum for the actualization of that cooperation.

The European Commission, as previously mentioned in Chapter 2, gathers national contact points for CIP under the official format of experts meetings, which are normally organized twice a year and in which the Republic of Croatia also participates. In these meetings, the Member States have the possibility to exchange best practices, discuss new threats and challenges (hybrid, cyber), and exchange their achievements in the protection of national and European critical infrastructures. For the Republic of Croatia, such a format of cooperation was very valuable in gaining insight on CIP practice in countries that have highly developed systems, and some of the good practices were implemented in our national CIP system. Also, meetings are held with external partners such as USA and Canada (EU–USA–Canada CIP expert meeting) and neighboring countries (EU–Neighboring Countries CIP Meeting). On meetings with neighboring countries[7], the representatives of the Republic of Croatia gave their suggestions on ways of future cooperation such as participation in thematic conferences (as already existing practice) and EU-funded CIP projects (which was also previously done with Serbia).

In addition to this formal network, the Republic of Croatia is actively participating in ERNCIP[8] meetings. This is an informal network, yet shared knowledge, expertise and solutions can be used for bettering the protection of European (and national) critical infrastructure from all types of threats and hazards. The Republic of Croatia is also using the possibilities of CIWIN as a cooperation tool for the exchange of information on strategies and risk minimization measures in the protection of critical infrastructure.

Also, the European Commission is cofinancing all interested actors in the field of critical infrastructure protection, and the Republic of Croatia has used that opportunity as well. The main purpose of project financing is to improve knowledge and provide a better understanding of the functioning of critical infrastructures (at all levels) and recommendations to public policies. In focusing on these areas of interest, the European Commission cofinanced project RECIPE 2015 for a dialogue between the Republic of Croatia, the Republic of Serbia and the Kingdom of Sweden. Through project activities (such as joint workshops, panel discussions, international scientific conference, follow-up strategy), there were few valuable project deliverables: feasibility study (to obtain the result of the most efficient and effective models for complete functioning of the entire CIP system in partner countries), "Resilience of Critical Infrastructure" guidelines (which include the best knowledge that was acquired during the project, representing a unique document which summarizes the best examples and suggestions of good solutions for the questions related to three main project topics: public-private partnership in the protection of critical infrastructure, challenges and mechanisms of sensitive information exchange among the stakeholders in critical infrastructure protection system, and setting preconditions for the development of national critical infrastructure centers), and mobile application "Recipe Network" (for getting together interested professionals involved in the protection of CI, providing with information about their area of expertise and competences for contacting them for business cooperation. The application also shares documents, project presentations and dissemination, events, etc.). Countries involved in the project had different stages of CIP development. The Republic of Croatia (which had some basic preconditions of development, a normative framework and initial activities) and Serbia (which just started with CIP system development, not having basics, and in that sense was the main beneficiary of this project) received guidance primarily from Sweden and other EU Member States on CI issues. It is very important to emphasize that the project has been very successful and has resulted in a multitude of long-term high-quality solutions that determined the needed direction of development of the CIP system in the Republic of Croatia.

From the presented, the option of using EU funds for CI-related projects, for cooperation through the European Commission and between the Member States and potential EU members is also recommended to all countries that are only at the beginning of CIP system development or want to improve existing ones.

3.5. Education and Training

In order for all stakeholders of the critical infrastructure protection system (state and private sector) to have the same basic knowledge for considering critical infrastructure issues and to carry out the necessary activities, it is necessary to establish certain education models through either theoretical education or practical application through trainings/exercises.

In the Republic of Croatia, in the period of the beginning of developing this area, there was no specific comprehensive program at any higher education institution[9] through which all who are engaged in activities related to CI could educate themselves on basic knowledge needed for better realization of duties and tasks related to the identifying, designating, protecting and strengthening the resilience of CI.

In pursuit of an ad hoc solution to provide a basic and the same level of understanding to Croatian experts (from representatives of the coordinator of the CIP system to security coordinators and their deputies from the nine ministries of relevant CI sectors) while harmonizing the expectations and knowledge of these experts, an educational seminar/workshop was created and conducted in 2014, entitled "Risk analysis of critical infrastructure operation". Following this seminar, there was an advanced course named "Mastering risk assessment and optimal risk management according to ISO 31000 and IEC 31010" in 2015. Upon completion of the course, participants took an exam prepared by the Professional Evaluation and Certification Board of Montreal, Canada (PECB) and received the following certification: Certified ISO 31000 Risk Manager. Both seminars were commissioned and financed by the NPRD and they were carried out by the University of Applied Sciences Velika Gorica in cooperation with external associates (Bluefield Company), which held lectures on risk management and accompanying standards ISO 31000 and IEC 31010 and with the use of the Hestia Risk Management software (an integral software solution for risk management using a qualitative methodology).

Although both seminars were quite useful and provided segmented knowledge, they could not be comprehensive enough, and moreover, they are not a regular activity. In the future, we should strive for systematic solutions, with continuous education. The courses were held only once, primarily because of finances, (ministries do not have the funds to participate in such education), so as mentioned, NPRD, as a national coordinator who wanted to support sectoral efforts in identifying critical infrastructure, has covered all the costs of participation. Finances should be provided within the competent public authorities in order to have additional training which can secure adequate implementation of the envisaged CIP activities.

This is closely linked to the strengthening of the role of the critical infrastructure coordinator, which could be improved by systematizing the specific workplace for them within the national authorities (which currently does not exist, and people designated as critical infrastructure coordinators also have their regular eight-hour jobs, usually not dealing with CIP tasks as one of working priorities).

In addition, it should be noted that in Croatia, the seminar in 2014 and the advanced seminar in 2015 were attended by people some of whom are no longer coordinators for critical infrastructure and did not transfer the acquired knowledge to their successors. The opportunities for education are limited and they need to be better aimed because otherwise, resources and knowledge will be "wasted". A possible solution for that issue is training the trainers.[10] Also, it is necessary to educate the owners/managers. This course would be of great importance for owners/managers because acquiring this knowledge is significant for achieving the functionality of critical infrastructures (such as ensuring business continuity).

Considering everything mentioned earlier, we can summarize that it is necessary to provide financial means, plans and education programs (as well as create adequate materials for education that will follow a clearly specified syllabus based on a multidisciplinary approach, guided by the principles of European regulation) for all stakeholders in the critical infrastructure (risk management) system – with an aim to increase knowledge and competences, include the scientific community, and impose an obligatory education. In all countries where the critical infrastructure protection system is highly developed, great attention is paid to education, so in the Republic of Croatia, it is also necessary to develop a model for the training of all key actors who have tasks and roles within the critical infrastructure protection system.

The Republic of Croatia has very good and long bilateral cooperation with Minnesota, USA (which started as military cooperation and became military-civil), and through that partnership, in January 2015, we had an opportunity to organize the "Identification, designation and protection of critical infrastructures" lecture held by representatives from the Minnesota National Guard and representatives of the Crisis Management Office of the cities of Saint Paul and Minneapolis. Since the Republic of Croatia had then just begun the process of identifying, designating and protecting national and European critical infrastructures, it was the opportunity to learn and share knowledge and opinions with those who are, for decades, the global leading experts in determining trends and standards in protecting and building the resilience of critical infrastructures. Lecturers were familiarized with the state of the process in the Republic of Croatia, and for this occasion, they were asked to provide basic principles and guidelines that represent the postulates of CIP. The lecture was attended by security coordinators from central state administration bodies, experts for protection and rescue activities from the Zagreb County and the city of Zagreb, as well as representatives of associations, institutions and other participants of the protection and rescue system.

In addition to education, trainings that are not only theoretical, but also give an opportunity to experience (and test) implementation of CIP system tasks, should be also carried out. Case studies and simulations are adding interaction to the whole process. In the Republic of Croatia, in line with new challenges for critical infrastructures, two staff exercises were organized: Cyber Shield 2018 and 2019, organized by the Co-ordination for the Homeland Security System and the Ministry of Defense. In 2018, there was a simulation exercise featuring cyber-attack scenarios (on water supply system, hospital and power system) with the national-level decision-makers gathered in the Homeland Security System Board. The objective was to raise the awareness of cyber threats and underline the importance of strategic communication in crisis situations, and the coordination and collaboration of the public, the academic, and the business sectors. Cyber Shield 2019 was a continuation of Cyber Shield 2018, but the difference was in focus on changes introduced with the adoption of the *Law on the Cyber Security of Key Service Operators and Digital Services Providers* (adopted in July 2018). During the exercise, the participants exercised their obligations and procedures proscribed by the mentioned Law, in case of a cyber-incident (in the scenario, it was Croatia Control (air traffic) and University Computing Centre as digital service providers). Cyber Shield is an example of good practice that should be more regularly conducted, involving more stakeholders of the CI system. Also, it would be valuable to organize such simulation (table-top) exercises with neighboring countries, having in mind interdependencies and EU critical infrastructures.

For all stakeholders in the CIP system, publications related to CIP can be a valuable addition to their efforts. From publications such as conference books of proceedings, "Critical Infrastructure in Republic of Croatia" (Matika & Poljanec-Borić, 2009) was for a long time a reference source, and it was created as a part of the interdisciplinary collaborative project "Energy Security and Critical Infrastructure", commissioned by Hrvatska elektroprivreda d.d.[11]), to various professional publications. One of the most recent publications is a book titled *Critical Infrastructure Systems: The Platforms for a Successful Development of Nations Security* (2018) which is the first comprehensive analysis in the Republic of Croatia on the concept of critical infrastructure – perceiving it from the aspect of the most developed countries of the world from which the concept of CIP origins (like USA), through EU development of the field (by the individual Member States and on Commission level) to the Croatian national efforts, presenting principles in the protection of critical infrastructures and strengthening of its resilience, using case studies and concrete suggestions how specified activities should be carried out in the Republic of Croatia and countries of the region (Mikac, Cesarec, & Larkin, 2018). The added value of this book is the interdisciplinary research that within the eight chapters covers several aspects of observation – scientifically, normatively, expertly, and also involving perspective of the public and private sectors and the applicability of certain offered solutions. It was well accepted among all experts in Croatia and became part of the curriculum in higher education institutions.

Chapter Conclusion

National critical infrastructures are facing growing threats – from natural disasters, asymmetric threats (such as terrorism, which "initiated" the development of the CIP concept) to cyber threats, and all of them are increasing the necessity of their protection. The key premise of successful critical infrastructure resilience and protection is that there is an established and effective ratio between normative solutions, implementing bodies and experts in the field – which makes an adequate system. It should primarily be focused on prevention and preparedness, cooperation at all levels in strengthening critical infrastructure resilience, and readiness for a quick response in case of an incident. If just one of these segments is not at the required level, the entire system may have essential problems in functioning or will not function at all.

States are already taking care of the infrastructure, which is of national importance, even if it is not defined under the concept of critical infrastructure – in the previously mentioned Croatian example, it is visible that the process of protection is very successfully carried out through a system of protection of objects particularly significant for the defense of the country. The transfer of the CI concept from the EU perspective should be aligned with the already existing national efforts, and for that, the key is the understanding of the process, political support and implementation synergy. While the CI is being perceived mostly from the aspect of national security, the perspective also extends to communities of countries (such as the European Union) and cross-border cooperation in creating a more resilient international network. New security challenges are coming, and critical infrastructure protection systems, regardless of individual country's development, need to be continuously upgraded and improved.

It is also clear from the practice that the establishment and maintenance of an efficient system seek immense energy, knowledge and continuous investment in material and immaterial resources, whereby the human factor (as the key to development) should be much more emphasized. The management of this system requires national strategic documents, action plans within each sector of the CI as well as the coordination of activities of all actors in the implementation of CIP. In spite of the fact that this is often challenging to achieve, all countries should aspire to that.

For the Republic of Croatia, the future of CI resilience and protection is very promising, with a great dynamic of its development. To achieve a stronger CI system, it is necessary to continue cooperation with other countries, different institutions and the private sector – as potential partners who could provide knowledge and best practices. For a more developed CI protection, it is also necessary to maintain a close relationship with the European Commission, using it as a platform which could give an additional impulse for continuing the development process. With regional and European cooperation in the protection of critical infrastructure, the "wider picture" is made, and that is very useful for improving the protection of the CI, namely, to help (and support) measures and activities on a national level to get closer to European positions.

Notes

1. This Strategy has been replaced with the new National Strategy for the Prevention and Countering of Terrorism (2015), which even more strongly underlines the importance of critical infrastructure as well as the need to build a national system and capabilities to protect the critical infrastructure and business continuity management (Government of the Republic of Croatia, 2015a).
2. Rules on criteria for designating and protecting objects particularly significant for the defense of the country define/describe objects that are particularly significant for a complete system of defense preparations and the creation of material reserves necessary for waging a war, namely, facilities for laboratory testing and production of weapons and military equipment and the most important objects on which continuous work in the war, the imminent threat of the independence and uniqueness of the Republic of Croatia, or in exceptional circumstances, largely depends the functioning of the economy and social activities aimed at supplying the armed forces of the Republic Croatian and other defense structures with the necessary products and services, and ensuring the basic needs of the population (Ministry of Defense, 1999).
3. These ministries are the Ministry of Economy, Entrepreneurship and Crafts, the Ministry of the Sea, Transport and Infrastructure, the Ministry of Health, the Ministry of Finance, the Ministry of Agriculture, the Ministry of Environment and Energy, the Ministry of Interior, the Ministry of Culture, and the Ministry of Science and Education.
4. As of January 1, 2019, the National Protection and Rescue Directorate has been integrated into the Ministry of Interior with the transfer of all competencies. NPRD was a CIP system coordinator from 2013 to 2019.
5. Croatian Telecom is part of Deutsche Telecom AG corporation and is the biggest of three telecommunication operators which are covering the entire national market.

6. Adria Pipelines (Jadranski naftovod – JANAF) is a public company that manages an oil pipeline and storage system, transporting crude oil and petroleum products.
7. The last one was held in 2018 with the objective of identifying possibilities for concrete cooperation projects between the EU and external partners in CIP. This meeting extended the cooperation further with eight partners (Albania, FYROM (Macedonia), Kosovo, Moldova, Montenegro, Serbia, Ukraine), and for the first time Georgia was included.
8. ERNCIP je JRC-facilitated network of security-related experts volunteering to address issues of prestandardization at the EU level toward fostering the development of innovative and competitive security solutions.
9. Although some of them like the University of Applied Sciences have courses, Critical Infrastructure (at Management in Crisis Situations professional undergraduate study program and Management of Logistic Systems and Processes specialist graduate study) and Critical Infrastructure Management (at Crisis Management specialist graduate study), they are not publicly available and are intended only for university students.
10. Educating a number of public sector employees with the necessary knowledge, which they would continuously transmit within the public and private sectors (of their respective competences).
11. A publicly owned Croatian electricity company.

EU ENLARGEMENT AGENDA

2

Turkey
Bosnia and Herzegovina
Serbia
Montenegro
North Macedonia
Albania
Kosova

DOI: 10.4324/9781003273769-5

Chapter 4

Turkey

Salih Bıçakcı and Ayhan Gücüyener Evren

Contents

Critical infrastructures (CIs) are major life-supporting systems and *sine qua non* of modern societies and cities since their proper functioning is essential for societal and economic well-being. In that respect, the unprecedented level of sophistication of cyberattacks, the increasing dependency on those interconnected and complex systems as well as their interdependent nature necessitates developing systemic perspectives and adequate policy responses. As being one of the OECD and NATO members as well as an EU candidate, Turkey is one of these countries whose functioning becomes increasingly dependent on CI systems.

According to World Bank's *Turkey Project* report, it is noted that "the main share of Turkey's GDP comes from the services (64%) and industries (27%), both which depend on CIs and lifeline utilities to preserve their competitiveness and productivity" (World Bank, 2018: 3). Thus, any significant damage and disruption in national CIs might directly affect economic and societal activities. Moreover, there are dependencies in energy and transportation domains between Turkish and European CIs such as the electricity line to Greece and two lines to Bulgaria

DOI: 10.4324/9781003273769-6

(European Commission Expert Group on Electricity Interconnection Targets, 2019) and natural gas pipeline projects such as Trans-Anatolian Natural Gas Pipeline (TANAP) and Turk Stream. Turkey is also a part of the world air traffic system that any hinder could easily affect the global service. Any protection vulnerability in the screening process of Turkey's airport would also create catastrophes for several destination airports. Ankara is also a key partner in other transportation means such as train and transit motorways which is critical for the logistics of various products.

During the last decade, Turkey implemented a number of critical strategic efforts particularly to strengthen the cybersecurity of CIs such as the implementation of successive National Cyber Security Strategies (2013–2014, 2016–2019, and 2020–2023), the national Cyber Incidents Response Team (CIRT – in Turkish: Ulusal Siber Olaylara Müdahele Merkezi – USOM) and mandate to establish Sectoral Cyber Incident Response Teams. Nevertheless, it should be noted that Turkey should consider initiating an independent, national, and resilient Critical Infrastructure Protection (CIP) governance methodology or a plan. This chapter aims to make an overview of the CIP landscape in Turkey by covering Turkey's normative framework and its relation to EU policies. Following the implementation of CIP policies, the public-private partnership (PPP) landscape, cooperation with the European Commission, and education and training efforts to consolidate and sustain CIP efforts will be handled.

4.1. Introduction: Normative Framework for Protecting CIs. The Efforts in Turkey

Each country's organization system, design of the CI, and bureaucracy determine how to strengthen and build resilient infrastructures. The connectedness of CI systems and design differences between these facilities hinder the change management and protection process. Two main practices come to the forefront in designing a broader CIP architecture: voluntary and mandatory approaches. While the voluntary approach consists of the "non-binding" guidance with the contribution and recommendation of all stakeholders, in the mandated approach, the establishment of a binding legal framework as well as implementing sanction mechanisms are the main tools to achieve CIP strategies (United Nations CTED and UNOCT, 2018).

Despite the increased interest and accelerated efforts particularly to secure cyberspace in Turkey, currently, there is no legislation that "directly" and specifically addresses CIP. Actually, CIP in Turkey is divided into two dissimilar domains: physical and cyber protection of the facilities. In this regard, some authors exemplify the *Statute 2011/2237 on Military Forbidden Zones and Security Zones* for the requirements of the physical security of energy, manufacturing, water management, transportation, telecommunications, intelligence, and military facilities (Karabacak, Ozkan Yildirim, & Baykal, 2016). Yet, this law is more focused on public services and gives less connotation to the private operators' responsibilities.

Another finding from the literature survey on CIP in Turkey is emerging interest in critical energy infrastructure (CEI) protection thanks to Turkey's attentiveness to industrial development and its connection to energy consumption. The cyber-attacks targeting Estonia, Georgia, and Iran as well as the Ukrainian case were surely influential to raise a concern about the security of energy infrastructures. Yet, according to some researchers, two specific incidents come to the forefront as wake-up calls for Turkish policy makers: the 2008 Baku–Tbilisi–Ceyhan (BTC) pipeline explosion and the 2015 countrywide blackout (Kasapoğlu, 2015). Even though both incidents were not officially confirmed as cyberattacks,[1] some national and official media reports blamed the Iranian hacker group Parastoo as responsible for the blackout that took place in March 2015 (ITProportal, 2015). Since energy security is a global concern and protection of CEI is an indispensable component in this particular calculation, Çelikpala assesses that "defining and securing Turkey's critical infrastructure including energy-related networks is now seen as a core part of nationwide legal, administrative and security system" (Çelikpala, 2009: 16).

This chapter exclusively aims to cover the normative efforts about CIP. In line with this effort, it cannot provide all the legislative framework about cyber security[2] even it is an indispensable component of securing CIs. Nevertheless, the Cabinet Decision No. 2012/3842 released in 2012 should be mentioned as a crucial milestone since it regulated for the first time the framework about "Implementation, Management and Coordination of National Cyber Security Efforts" (Bıçakcı, Ergun, & Çelikpala, 2015). As a result of this decision, the Cyber Security Board was designated as the top governmental organization regarding the governance of national cyber security (Şeker & Tolga, 2018). Various public institutions were also represented in this board to communicate and participate in this structure.

While the Cabinet Decision of 2012 set forth preparation the policies, strategies, and action plans to provide the National Cyber Security, under the coordination of the Ministry of Transport and Infrastructure, three successive national cyber security strategy plans were prepared and released for the periods 2013–2014, 2016–2019, and 2020–2023. All National Cyber Security Strategy documents put a great emphasis on CIP and formed a subcommittee focused on CIP in the Cyber Security Board to determine the CI sectors. The subcommittee decided on six CI sectors (Electronic Communication, Energy, Water Management, Critical Public Services, Transportation and Banking and Finance) while defining CIs as "infrastructures that contain information systems, which may cause loss of life, large scale economic loss, national security gaps or disturbance of public order when the confidentiality, integrity or availability of the data they contain is compromised" (Ministry of Transport and Infrastructure, 2016). When we assess 2016–2019 strategy document's specific efforts to secure CIs, "creation of a national critical infrastructure inventory", "meeting security requirements of critical infrastructures" and "supervision of these critical infrastructures by the relevant regulatory board" came to forefront as important targets and milestones (Ministry of Transport and

Infrastructure, 2016: 26). Even though the provisioned timeline for the 2016–2019 strategy is expired with the announcement of the latest 2020–2023 strategy, it is not publicly declared whether all of these objectives are fully achieved since the national strategy is not a legally binding document. According to the 2020–2023 strategy's assessment of the previous period, it is noted that "monitoring of critical infrastructures service continuity, vulnerability scanning, regulation and supervision studies in terms of work and information security was carried out" (Ministry of Transport and Infrastructure, 2020: 19).

Turkey's last National Cyber Security Strategy and Action Plan for 2020–2023 was issued in December 2020 (Ministry of Transport and Infrastructure, 2020). Similar to the previous strategy, securing CIs is identified as one of the prominent strategic targets. Yet, different than the previous strategies, the "resilience" concept has been included in the strategy for the first time. Also, "24/7 protection of cyber security of CIs", developing an approach based on "regulation and supervision" of CI sectors, and "eliminating vendor dependence for IT equipment in CIs" are noted as strategic targets to achieve for the forthcoming years.

In terms of cyber security, some CI sectors, such as energy, are better regulated and coordinated in Turkey due to the presence of powerful regulatory agencies, influential sectorial NGOs, or coordinated private sector efforts. For instance, the Energy Market Regulatory Authority of Turkey (EMRA – in Turkish: Enerji Piyasası Düzenleme Kurulu – EPDK[3]) has announced "Regulation on Information Security of Industrial Systems Used in the Energy Sector" in July 2017, which proclaims that CEI operators[4] should "prepare a risk inventory to monitor the information process and ensure safety of industrial control systems used in critical energy infrastructure" and "prepare a treatment plan clearly outlining risk mitigation actions" (Official Gazette of Turkey, 2017).

When the recent normative landscape to secure CIs in Turkey is examined, we can discover that the *coup attempt* was influential to expedite the concerns arising from insecurities in cyberspace. Even though a comprehensive national cyber security framework was already launched with the implementation of national cyber security action plans (i.e. identification of national CIs), the institutional reforms under the Presidential system triggered a centralization trend in national cyber security governance aiming to unite all distributed cyber security tasks under Digital Transformation Office of the Presidency of the Republic of Turkey.

While a strong commitment to creating a single cyber security leadership and a consolidated national cyber security framework is significant, the *Presidential Circular on Information and Communication Security Measures* released in July 2019 could be listed as a significant outcome of this process as it aims to "reduce security risks and promoting governing measures to be taken to ensure safety of information which are critical to national security and public order" (Official Gazette of Turkey, 2019). Additionally, while the circular imposes several security obligations for public entities and CI operators, it has also remarkable consequences for the CIP regime in Turkey. Some of the important points in the circular concerning security of CIs but not limited to are as follows:

- Obligations for secure storing of critical information (i.e. population, health, contact, and biometric information).
- Isolation of industrial control systems (ICS) from the Internet; if the connection is necessary, deploying required security measures such as authentication and authorization.
- Conducting personnel security investigation for the personnel who will be employed in strategic institutions critically important for national security.
- According to the circular, the responsible institutions shall establish inspection mechanisms for the fulfillment of these obligations and conduct an audit at least yearly. Additionally, the report which covers the results of the inspection will be transmitted to the Digital Transformation Office.

As underlined in the circular, *Information and Communication Security Guide* is launched in July 2020 (Digital Transformation Office of the Presidency of the Republic of Turkey, 2020a). With the extensive participation of different stakeholders, this Guide serves as the first national document which imposes the obligatory security measures to be taken by public institutions and CI operators which are using information technologies. As noted in the document, while authenticity is the primary principle, the content of Guide is prepared by also taking the national/international best practices and standards (i.e. NIST Special Publication 800–53, OWASP IoT Security Guidance) into consideration. In the Guide, the security of CIs is handled as a specific subchapter (Section 4.5). In this chapter, in addition to the principal security measures (i.e. network and system security, IoT devices' security, security of physical spaces), in-depth security measures are particularly listed for two specific critical sectors: energy and electronic communication. Some of these obligatory measures for the energy sector are demonstrated in Figure 4.1. Finally, along with the announcement of the circular and the guide, anticipated *Cyber Security Law* will be the main components of normative framework about the cyber security of CIs in Turkey.

Energy sector	Preventing unauthorized access and providing access control for the communication between ICS network and corporate IT network
	Using separate authentication systems for ICS users and corporate users
	Defending GSP technologies deployed in energy infrastructures against spoofing attacks

Figure 4.1 Some examples of security measures for energy sector in the information and communication security guide

4.2. Implementation

The threats of the 21st century are so dynamic that any CIP governance or framework should have an outstanding adaptation and flexibility capacity to respond to the emerging threat landscape. This is only possible if a fostered cooperation between public-private actors takes place and proper research and development (R&D) efforts are implemented.

When we elaborate on the implementation of CIP strategies, the national governance, structure, and related stakeholders can be analyzed. In Turkey, there is no supreme authority whose tasks are solely dedicated to CIP and even there is no national CERT dedicated to the cyber security of CIs. As mentioned in the normative framework, in terms of cyber security of Critical Information Infrastructures (CII), along with the supervision of the Ministry of Transport and Infrastructure, a Cyber Security Board was assigned as being the main governmental authority to implement national cybersecurity strategy, policy, and roadmaps. Yet, in line with the recent centralization approaches in the national cyber security framework, the Digital Transformation Office under the Presidency gains remarkable authority on cyber security of the CIs. In that sense, the Cyber Security Department under Digital Transformation Office is appointed with responsibilities of "developing cyber security strategies for CIs", "carry out studies to identify CIs", "supporting national and domestic solution/product development activities to ensure widespread usage in CIs", "carrying on preventive and protective activities to secure critical technologies and informative assets" etc. (Digital Transformation Office of the Presidency of the Republic of Turkey, 2020b). Yet, some dimensions regarding which other authorities lead the necessary technical preparations and which mechanisms will be used to enhance PPP still remain unclear.

Along with Digital Transformation Office, some of the other key institutions like the Scientific and Technological Research Council of Turkey (TÜBİTAK), Turkish Information and Communication Technologies Authority (BTK) and the Turkish National Computer Emergency Response Team (USOM) shall be cited since they have vital roles in national CIP efforts. TÜBİTAK shows noticeable efforts both in developing guidelines and best practices and in supporting national technologies to strengthen Turkey's CIP landscape. For instance, TÜBİTAK has prepared the *Minimum Security Measures Document for Critical Information Infrastructure* which is served as a roadmap for CI operators on the protection of systems from unauthorized access, access management of authorized personnel, management of system supply, development and maintenance, business continuity measures, employment of information security systems, security manager and employees, documentation, and information security incident responses (TÜBİTAK, 2017). This guideline attributes special importance to Operational Technologies (OT) and defines CIs as "systems which comprises information systems or ICS that would cause loss of life, large-scale economic damage, national vulnerabilities or disturbance of public order in case of the loss of privacy, integrity or accessibility of the information that they processes" (TÜBİTAK, 2017: 6). In spite of these guidelines, these institutions are offering limited information to the sectoral operators to implement cyber security policies and their improvement.

USOM is established under BTK and ensures national and international coordination for the prevention of cyberattacks against critical sectors (Şeker & Tolga, 2018). In addition to USOM, sectoral SOMEs (CERTs-in Turkish: Sektörel Siber Olaylara Müdahale Ekipleri – Sectoral) are other key stakeholders in Turkey's CIP implementation since the *Communique on Principles and Methods regarding Establishment, Tasks and Works of CERTs* obliges to set up a SOME in all CI sectors defined in the National Cyber Security Strategy Document (Official Gazette of Turkey, 2013). To facilitate this process, the Ministry of Transport and Infrastructure prepared the *Guide for Creating and Managing National Cyber Incident Response Team* (Ministry of Transport and Infrastructure, 2014).

Even though cyber security is a vital component of CIP policies, securing CIs requires a broader perspective that integrates the *All-Hazards* approach since "CIs are subject to polymorphous types of threats" (United Nations CTED and UNOCT, 2018: 24) including natural disasters or unintentional human errors. For Turkey, OECD's factsheet also makes a distinction between natural hazards and man-made risks and states that "Turkey is exposed geophysical hazards particularly earthquakes, as well as industrial accidents, cyberattacks and organized crime pose a significant threat" (OECD, 2017: 100). Yet, in Turkey, the term "critical infrastructures" is generally used to address "critical information infrastructures" since most of the national efforts are concentrated on strengthening the national cyber security framework.

In terms of competent institutions which are dealing with securing CIs, we can firstly mention the role of Disaster and Emergency Management Authority's (AFAD – in Turkish: Afet ve Acil Durum Yönetimi Başkanlığı) role. The main motivation behind founding the Authority has firstly emerged after the 1999 İzmit earthquake which showed the necessity for establishing a governmental institution to single-handedly coordinate and exercise legal authority in cases of disasters and emergencies (AFAD, 2019). In 2009, crisis management and CIP have been delegated to AFAD with the Law No. 5902 (Bıçakcı, Ergun, & Çelikpala, 2015) Additionally, the institution is also delegated with the responsibility to coordinate preparedness and post-disaster response and to promote the cooperation among various government agencies (OECD, 2017). AFAD shows a noticeable performance in humanitarian assistance as well as mitigation of disasters and emergencies (particularly related to non-man-made risks and threats). While the institution was functioning under the Office of the Prime Minister until 2018, the Authority now serves under the MoI which shows a more "internal security" focused approach rather than just coping with disasters (Ak, 2019).

As being the designated authority, AFAD also prepared for the first time a roadmap for CIP in Turkey. While *2014–2023 CIP Roadmap Document* categorizes disasters as "natural" and "technological[5]" ones, it also lists several necessary action items related to CIP to be finalized such as: "Preparing draft regulations concerning harmonization with European Union directives", "creating and implementing training programs", and "preparing a national plan for CIP to safeguard CIs" etc. (AFAD, 2014: 42–46).

When it is compared with other national best CIP practices, there are some gaps to be addressed in Turkey's CIP regime. For instance, UN's Compendium of Best Practices on CIP assesses that "whatever institutional model is chosen, stakeholders involved in CIP need to be familiar with this concepts and consistently apply them within their respective sector and compliance" (United Nations CTED and UNOCT, 2018: 50). Yet, the designated authority in Turkey, AFAD, is not an institution whose primary expertise is cyber security or protection against physical attacks. Besides, the existing roadmap is not clear enough about how cyber security crises will be managed (Bıçakcı, Ergun, & Çelikpala, 2015) and a comprehensive national plan on CIP is still missing. In terms of physical security of CIs, while the main responsible central body is the Turkish MoI (i.e. in rural areas dams and refineries are protected by Gendarmerie General Command) (Çelikpala, 2013), maintaining cyber security is considered an operator's task as these differentiated responsibilities would create a fragmentated environment for CIP policies' implementation.

Additionally, security assessments that focus on only one aspect of security might be ineffective (Pollet & Cummins, 2009). OECD emphasises that the "all-hazard and threat approach to critical infrastructure resilience and security enables policy makers and operators to better prepare for the unexpected" (OECD, 2018: 2). From this perspective, one of the other main problems in Turkey's CIP practices is about how to approach this field since the current national structures and strategies are mostly handling CIP as a subchapter of cyber security (Afyonluoğlu, 2017). Also, the existing efforts are concentrating on prevention, detection, and mitigation of the risks and threats. The emphasis on "resilience" is almost missing although this critical concept is an indispensable part of CIP practices and it is related to "focusing on the capacity of systems to adapt, reorganize and recover from disruption and disturbance" (Zio, 2018). While the resilience concept is limited in the existing Turkish strategies, cultivating a common understanding of the resiliency of CIs might be a good starting point. Moreover, this principle should be integrated into multi-hazard management (IBRD & World Bank, 2018).

4.3. Public-Private Partnership

Neither public authorities nor the private sector agents have the sufficient resources, budget, and expertise to secure the CIs alone. Thus, the need for PPPs has become the "mantra" for the CIP community as this specialized relation is defined as "contractual cooperation between public authorities and private-law organizations"

(OSCE, 2013: 69). Even though most of the advanced CIP practices use PPP mechanisms as the main tool, there is no universal scenario or receipt about setting a successful PPP model. Cultural differences and characteristics are important in setting different practices. ENISA also recognizes the shaping force of culture in building PPP models in EU states and claims that "it is evident that culture is the most important determinants of the way PPP are being established, develop and work" (ENISA, 2017: 18).

The PPP models also might vary from being project-oriented and process-oriented to institutionalized PPPs. Also, the factors such as ownership of the infrastructure (public versus private), regulations within a sector, anticipated threats, and preferred protection actions (i.e. reducing the consequences versus rapid recovery) shape the preferences about which PPP model would be implemented (CISA, 2019). While the CIP community arrived at the consensus on the indispensable role of the private sector in ensuring resilience of CIs, DunnCavelty and Suter discuss that "CIP policy should increasingly rest on self-regulating and self-organizing networks, the government's role would no longer consist in directing and monitoring, but of coordinating the networks and identifying instruments that can help motivate networks" (Dunn-Cavelty & Suter, 2009: 179). Yet, lack of trust among stakeholders or insufficient human resources come to the forefront as the main challenges. In that sense, senior leadership support, leadership in the partnership, institutional design, incentives and value, timing, urgency, and maturity are listed as five key items which contribute to a PPP's proper functioning (CSIS, 2013).

While privatization of infrastructures is a rising trend in Turkey, according to investment office, Turkey has implemented $139 billion worth of PPP projects in a variety of sectors (Presidency of the Republic of Turkey Investment Office, 2019). Energy is one of the most prominent sectors in which the private sector's role is rapidly increasing. For instance, while the private sector has a 57% stake in electricity production in 2003, this rate accelerated to 71.3% by 2014 (Hürriyet, 2014). The same privatization trend is also present for other critical sectors like health. In that sense, while the number of hospitals reached 1,534 in 2018, 577 of them have been operated by the private sector (Gürtaş, 2019). In such context, when the Turkish CIP approach in terms of building PPPs is examined, significant steps were taken to ensure engagement of the private sector. However, an institutionalized/specialized PPP that focuses on CIP in a holistic way is still absent in Turkey.

In the early efforts of creating the Turkish cyber security governance framework, the participation of private sector agents was almost negligible since private actors were not even listed among the members of the former Cyber Security Board. For instance, the EMRA, which is also nominated as the sectoral SOME for the energy sector, was not represented in the Board (Cybermag, 2018). Yet, under the ongoing institutional transformation process, it is expected that the engagement of the private sector and the public-private sector coordination would accelerate. The recently established Digital Transformation Office pays special attention to expanding the inclusiveness of the national cyber security framework. The Turkish cyber security ecosystem is also strengthening under the "Turkish Cyber Security Cluster" which is founded in order to "develop Turkish cyber security ecosystem with the contribution of all public agencies, academia and private sector representatives by leadership of Presidency of Defence Industries" (Türkiye Siber Güvenlik Kümelenmesi, 2020).

Yet, most of the existing efforts are concentrating on cyber security of Cis, and they are not approaching CIP from a broader and *All-Hazard* perspective. This is particularly related to the absence of a central institutional body like the US Cybersecurity and Infrastructure Security Agency (CISA), which is responsible for "leading the National effort to understand and manage cyber and physical risks" to CIs. In terms of the cyber threat information sharing across different stakeholders, Turkish CIRT's (USOM) involvement is significant since the institution is tasked with constantly monitoring and providing warnings and announcements for cyber security incidents (Şeker & Tolga, 2018).

To sum up, mobilizing PPPs and building threat information and intelligence sharing mechanisms are generally problematic and sometimes ambivalent. Private sector partners are sometimes reluctant to share information due to the "concerns about liability, resources cost, loss of proprietary information and return of investment" (Fleming, Goldstein, & Roman, 2014: 7). In some cases, the public authorities are also unwilling to share the information with the necessary recipients due to their organizational schemes which are not designed to distribute information regarding cyber security threats. Nevertheless, with respect to the best practices in CIP models, Turkey can consider building more institutional threat information sharing bodies for national CIs which would further clarify the lacuna related to "who should share information", "what should be shared", and "when should the information is shared". Sector-specific Information Sharing and Analysis Center (ISAC) could be considered as a reference model in building such an institutional framework for Turkish CIP practices and community. More importantly, the primary step in this process should be launching a national CIP plan which would set strategic and measurable targets with strong leadership, clear tasks, and responsibilities.

4.4. Cooperation with European Commission

Despite ups and downs in relations, Turkey obtained the candidate country status in 1999, and the European Council decided that Turkey sufficiently fulfills the criteria to open accession negotiations in December 2004 (Council of the European Union, 2020). In that sense, as a candidate country, adaptation and alignment of most of Turkey's national legislation in line with the EU law is necessary. Throughout this process, the CIP topic is handled with the "civil protection" part which takes place under the Environment Chapter of accession negotiations (AFAD, 2014: 20). When it comes to Turkey's cooperation with the EU on CIP, it is worth mentioning AFAD's CIP roadmap since the document is prepared as a guideline to align Turkish CIP policies' implementation with those of the Union. In that sense, the roadmap suggests a *to-do list* for CIP practices in Turkey which includes preparation of a draft bill for alignment with EU regulations, closer cooperation with relevant EU stakeholders, and integration of Turkey's practices with EU's CIWIN (AFAD, 2014). Thus far, most of these action items have not yet materialized.

Regarding the cooperation with the EU, we also need to mention the current regulatory efforts in Turkey to align with the EU's NIS Directive. As a candidate country, Turkey conducts several projects to align its regulatory framework and institutional capacity with the NIS Directive. As a result, in 2019 *Technical Assistance Project for Harmonization with the EU Regarding the Implementation of the Network and Information Security Directive* has launched. In line with this project, gap analysis report and legislative proposals were recently prepared (KMO Information Systems, 2020). It is expected that these intensified legislative and regulatory efforts on cyber security on CIs will be handled and released with a comprehensive cyber security law since Turkish authorities have been working on such a regulation package.

The EU attaches great importance to R&D efforts to secure CIs. Thus, the Commission recently announced that it has committed more than €38 million through Horizon 2020 Programme to support innovative projects in the field of CIP against cyber and physical threats and for making cities smarter and safer (European Commission, 2020). Turkey has been a part of European research framework programs since 2003. The ongoing partnership under Horizon 2020 and Horizon Europe (for the period after 2021) are also significant components of Turkey-EU cooperation on CIP, particularly, in terms of joint R&D efforts as well as human capital and knowledge exchange.

4.5. Education and Training

Securing CIs is a process that is embedded in a national security culture formulated by various factors such as national experiences, history, and threat perceptions. Adapting this security culture to today's hybrid and converged threats is only possible if relevant individual and organizational expertise is properly built as well as a strong effort to train the necessary workforce is set forth.

CIs are complex socio-technical assets. Given their sizes, complexities, and interdependencies, training and education efforts for securing them require a multidisciplinary approach that should involve different engineering disciplines as well as different branches of social sciences. Yet, in most of the countries, including Turkey, the number and level of academic as well as professional CIP course programs and curriculums are insufficient. One of the biggest challenges which impedes the development of widespread educational programs and the emergence of CIP as a formal field of study is an underestimation and lack of interest since most of the existing efforts have not yet captured educators' and organizations' attentions (Oliver & Haney, 2017).

As indicated in the previous chapters, the current legislative and structural efforts in CIP in Turkey are mostly focusing on the cyber security of CIs. The same argument is valid for education and training efforts for CIP in the Turkish model. While the number of universities that are providing graduate-level cyber security courses is increasing, some Turkish universities also launched cyber security (and critical infrastructure protection) research centers and laboratories.[6] Additionally, in 2019, Turkey's Defense Technologies Engineering and Trade Inc (STM) and Sakarya University have launched a project to build a "national test bed for CIs based on Industrial Control Systems" which is representing a university-industry partnership initiative for CIP. The State Supervisory Council's cyber security report recommended decision-makers to launch cyber security at the elementary school level to increase societal cyber vigilance (Milliyet, 2018). Turkish private cyber security companies and some state-supported institutions started to develop cyber security curriculums and establish academies to support human capital development and training.

In addition to the academic degrees and programs, exercises and competitions organized at the national level are other prominent tools for educating and attracting young talents. For instance, while STM is established to provide system engineering, technical support, project management, technology transfer, and logistics support services for the Turkish Armed Forces (TAF), the Undersecretariat for Defence Industries (SSM) organizes "Capture the Flag (CTF)" competition annually to encourage talented students and professionals to pursue a career in cyber security (STM, 2019), BTK organizes "Siber Yıldızlar" (Cyber Stars) contest. It is reported that there were 3174 participants for 2019's Siber Star competition and some of the successful participants were hired by Turkish CERT, USOM (BT Haber, 2020). In terms of cyber security of CIs, in September 2020, Turkey hosted a specific international CTF competition called "HackZeugma" which has focused on different threat scenarios based on cyber security of ICS.

Turkey shows a noticeable performance in developing a strong national cyber security framework. In that respect, the country's rating in the 2018 International Telecommunication Union's Global Cyber Security Index (ITU GCI) remarkably improved in both global and European levels in comparison to the previous years. But, since most of the efforts focus purely on the technological perspective of cyber security, Turkey needs to consider establishing training and education programs that consider accidental, intentional natural threats to CI in a coordinated manner. According to best practices suggested by the UN, "the all-hazards approach is often seen as a prerequisite to make the best use of limited resources and avoid needless duplication" (United Nations CTED and UNOCT, 2018: 24). Thus, nominated CI associations/regulators – like EMRA in the energy sector – should deliberate to launch sector-specific CIP training/education programs encompassing prevention and mitigation of all types of threats and enhancing the resilience of CIs. Moreover, the involvement of academia is certainly important for building a consolidated national education program but the practical dimension and hands-on experience should be integrated into the existing practices.

Chapter Conclusion

Turkey is taking steps forward to secure its CIs. However, these steps and efforts are always behind the technological improvement and private sectors' ambition to keep up with the technology as in the case of several countries. The major role of the national CIP bodies is to motivate the private sector to pursue its cyber security programs. The coordination responsibility of the state should not be ignored in case of an emergency. The coordination task comes with prerequisites such as strategic communication, intelligence, transparent governance. The CI operators are mostly looking for a mentor than a protective shield that would assist them in case of emergency or a Sherpa that would protect them from the sectoral threat. In the last decade, Turkey has come a long way in the CIP especially from the legislative perspective. However, it is still far from establishing or implementing a total awareness in CIP with the contribution of all partners at the national level.

The models for the protection of CIs implemented worldwide are swiftly changing due to the emerging and hybrid threats and vulnerabilities. Moreover, performing these national CIP practices requires continuous effort and diligent improvements to have remarkable success. In such context, Turkey's effort in CIP requires a smart and dynamic orchestration to protect its assets. The panorama of CIP in Turkey formed up with several sectors and their sub-branches. The quantity of all these bodies impedes easy and quick solutions for the security of the sector. The scale of the sector also amplifies the components and areas associated with the security policies. To improve a holistic approach, the government should make projects and plans to implement such understanding which is accessible to all participants of the national CIP setting. While the Turkish CIP practices come to the forefront with its attentiveness to strengthen the national cyber security posture, the government might nominate a specialized agency that will approach CIP from the *All-Hazards* perspective and ultimately coordinate all CIP-related activities nationwide.

In many countries, technology is changing faster than regulators. As a result, most of the existing CIP efforts for critical services and assets are designed to respond to the former expectations and remain far beyond to meet the requirements of the modern threat scene. In this respect, Ankara should continue to perform its intensive efforts but embrace a new mentality that permits a flexible design and less hierarchical organizational structure that focuses more on the holistic protection of the CIs. In such a structure, strong and mutually beneficial partnerships between public-private agencies as well as operators and the government are strongly recommended. The government might consider putting *ad hoc* or *structured* partnership mechanisms in practice which would not only accelerate the trust among involved actors but also ensure accountability as well as facilitate monitoring the activities. Last but not the least, there should be an open and transparent communication between the public- and private-sectors, and they should synchronize as much as possible. On regular days, the sector should get used to communicating with the national CIRTs to overcome all their problems. This also requires a particular CIRT that has remarkable expertise on SCADA and other OT types. Not only Turkish CIRT but also the global sector faces a remarkable crisis to find the required human capital with such background. Therefore, the Turkish Cyber Security Cluster should establish a plan to raise such talents to support the CIs. Additionally, efforts to establish a national ICS testbed laboratory should be accelerated and expanded to access all CIP stakeholders to test and approve trustability of the critical equipment which are essential for the functioning of the CIs.

Finally, with respect to the transboundary nature of critical assets (i.e. pipelines, power grids, transport networks) as well as Turkey's key location in providing and hosting crucial services and facilities, coordinating national CIP efforts and resilience policies with neighboring countries and regions is a key requirement. Turkey should also go over its emergency case scenarios with NATO or the EU to test their cooperation capabilities. All parties should also focus on building resilient infrastructures. Maintaining a resilient system depends primarily on "people" rather than technology. In all these efforts, strengthening the national human capital in Turkey is key to sustaining high-quality and vivid protection efforts not only for the public services but for the private operators as well. CIP is not a project but a process which has to continue without any signs of fatigue.

Notes

1. The Final Report, concerning the blackout which occurred on 31 March 2015 in Turkey, is prepared by Project Group Turkey and is issued by Turkish Electricity Transmission Corporation (TEIAS) and European Network of Transmission System Operators for Electricity (ENTSO-E) based on information available. Access: https://eepublicdownloads.blob.core.windows.net/public-cdn-container/clean-documents/SOC%20documents/Regional_Groups_Continental_Europe/20150921_Black_Out_Report_v10_w.pdf, (Access date: 12.05.2020).
2. Cyber security is not regulated by a particular legislation, but it can be found in other regulations. In terms of the legislative dimension of Turkey's cyber security efforts, some crucial steps could be introduced: Cybercrimes were first introduced to the

Turkish penal system on 6 June 1991 with Law No. 3756 and Turkish Penal Code's article Nos. 243, 244, and 245 are focusing on cybercrimes. In addition, the Law on Security of Electronic Communication is another instrument about regulation related to cybersecurity. Law No. 5651 on Regulation of Publication on the Internet and combatting crimes committed by means of such publication is also somehow relevant to CIP in Turkey due to its patronage of ISP controls.

3. EPDK – Enerji Piyasası Düzenleme Kurumu was established in 2001 in order to perform regulatory and supervisory functions in the energy markets.
4. Electricity transmission license holders, electricity distribution license holders, electricity generation facility owners that have temporary acceptance and installed power of 100 MW or more, natural gas transmission license holders which undertake transmission via pipeline, natural gas distribution license holders which are obliged to establish a shipping control center, natural gas storage license holders (LNG, underground storage), crude oil transmission license holders, refinery license holders.
5. According to the roadmap, technological disasters were defined as disasters and emergencies that would lead to loss of life or property and triggered by human activities or natural events such as industrial, mining-related, transportation, logistics, radiological, and nuclear incidents, incidents that cause sea pollution, large fires, and cyber incidents. Even though cyber incidents seem to be included in AFAD's roadmap as a technological disaster, rather than AFAD, generally, BTK is involved in the mitigation of the cyber threats.
6. For instance, Kadir Has University's Center for Cybersecurity and Critical Infrastructure Protection (CCIP) is one of the first established centers focusing on CIP research, such as security of industrial control systems, energy, and other critical infrastructures. Access: https://ccip.khas.edu.tr/9/about.

Chapter 5

Bosnia and Herzegovina

Mirza Smajić

Contents

This chapter offers an analytical overview of the situation as to the development of the system of the critical infrastructure protection in BiH. Furthermore, the chapter includes the analysis of the level reached in building a normative and strategic framework of the critical infrastructure in the State. The analysis shows that normative and strategic documents in the area of civil protection carry certain advantages that may be used to establish the critical infrastructure protection system. Still, BiH, due to its political and social, as well as security-related peculiarities, has not dedicated sufficient attention to the protection of the critical infrastructure. The consequence is that the Law on Critical Infrastructure and other bylaws regulating this area have not been passed. Certainly, the chapter provides certain guidelines aimed at developing the starting points for the critical infrastructure protection system in BiH, which stems from the obligations that have to be met on the path toward full membership in the EU and NATO.

DOI: 10.4324/9781003273769-7

5.1. Introduction

The main foreign policy priorities and objectives of the State of Bosnia and Herzegovina include the integration into the European Union and NATO. However, there are certain peculiarities when it comes to the State of Bosnia and Herzegovina, which are the product of the General Framework Agreement for Peace[1], and the functioning of BiH's security system depends, primarily, on internal politics as well as on international issues and relations. In practical terms, the Entities/Cantons are the primary institutional holders of the internal security system, as well as of other elements relating to civil protection, or to critical infrastructure protection. In that context, BiH has partially started the reform of the security sector, as well as of the civil protection sector. Still, due to internal social and political animosities, which constantly generate an "administrative vacuum", the expected results have failed to materialize or have been imperceptible, with the final result being the standstill on the road of the transformation of the society and the State and its integration into the Euro-Atlantic security structures. The state orientation, when it comes to the mechanism of CIP, has not been implemented yet as a whole, but has been developing through certain elements and mechanisms. Here we refer, first and foremost, to the adoption of certain normative and legal and strategic documents, as well as legal solutions relating to the area of CIP (*Framework Law on the Protection and Rescue of People and Material Goods from Natural and other Disasters in Bosnia and Herzegovina* (2008); *Risk assessment for Bosnia and Herzegovina regarding natural and other disasters* (2011); *Strategy for prevention and fight against terrorism* (2013); *Law on Security of Critical Infrastructures in Republic Srpska* (2019)).

Over the recent years, the issue of establishment or the reason for the lack of the framework for the critical infrastructure protection in BiH has become an inevitable topic in the public, media and academic space. However, a significant and visible move in the implementation of certain steps aimed at establishing the framework for the critical infrastructure protection pursuant to the European and other regulations (NATO) and standards, regretfully, has not been established as yet.

5.2. Implementation

From our point of view, BiH has not yet built a complete and coherent system to regulate separately the critical infrastructure protection. The fundamental document that regulates the issues and standpoints on the security of the state was adopted in 2006 under the title *Security Policy of Bosnia and Herzegovina* (Presidency of Bosnia and Herzegovina, 2006). This document did not define the area of CIP, it did, however, certainly create the preconditions for the development of other institutional and normative capacities that may be correlated to CIP. In that context, while bearing in mind the foreign policy orientation of the State of BiH, one of the fundamental security interests of the State of Bosnia and Herzegovina is the "approximation and institutionalization of relations with important international structures, including joining the European Union and the NATO Alliance" (ibid., 2006: 17–18).

The first visible progress that may be associated with certain elements of the system of CIP is contained in the *Methodology for the preparation of Risk assessment of Bosnia and Herzegovina regarding natural and other disasters*, i.e. the principal objectives of the preparation of the risk assessment include the "assessment of exposure of people, property, critical infrastructure to all major hazards" (Council of Ministers of Bosnia and Herzegovina, 2009). On the basis of the mentioned document, the preparation of the *Risk assessment for Bosnia and Herzegovina regarding natural and other disasters* was commenced and it was finalized in 2011. This document offers an overview and analysis of all possible hazards and risks that are a threat to BiH, as well as assessed needs and possibilities to prevent, reduce and eliminate the consequences of accidents and (or) disasters (Council of Ministers of Bosnia and Herzegovina, 2011). Although the mentioned document comprises certain parts of critical infrastructure, or hazards that may be a threat (hazards in transport and communication; Industrial hazards and such like), still a complete definition to include all parameters for the designation of critical infrastructure has not been defined. Along these lines, the State of Bosnia and Herzegovina has not started the process of building the critical infrastructure system. Here, the particular emphasis is placed on the necessity for the state structures to start building the system of the critical infrastructure protection. That process should take place, considering, in particular, the European aspirations, in accordance with the *Council Directive 2008/114/EC*, which sees the implementation of the critical infrastructure protection in three stages: (1) the identification of potential critical infrastructure; (2) the designation of critical infrastructure and (3) the critical infrastructure protection (Council of the European Union, 2008).

What is more, the adoption of strategic documents in the area of prevention and fight against terrorism proved that the relevant political actors should create an adequate system of the critical infrastructure protection in BiH. Along these lines, the *Strategy of Bosnia and Herzegovina for prevention and fight against terrorism in the period of 2010–2013* clearly identified the necessity to define in the form of law the critical infrastructure and the plan of protection and to establish the "early warning system" (Ministry of Security of BiH, 2010: 11–27). In addition, as far back as 2009, BiH opened within the Ministry of Security of BiH a modern operative-communication center (112) with a view to collecting, analyzing and distributing data, as well as coordinating activities during civil protection. However, due to certain internal political relations on the one hand, and the lack of normative and legal regulations on the other, the center has not received and served its full purpose and function in the civil protection system of BiH. Further, the currently valid *Strategy for prevention and fight against terrorism (2015–2020)* shows that the issues of the establishment of fundamental elements of the system of the critical infrastructure protection in BiH are still not in the focus of the political and social public. One of the fundamental sub-objectives of that document pertains to the critical infrastructure protection, namely with the aim of efficient implementation of the set measures, the emphasis is placed on the obligation to adopt "minimum standards for the critical infrastructure protection" in BiH (Ministry of Security of BiH, 2015: 7–9).

That the State of Bosnia and Herzegovina has to start creating the minimum standards for the critical infrastructure protection also follows from the fact that

> contemporary threats today are increasingly aimed at endangering economic activities and, in that context, the critical infrastructure is a potential target . . . States have the obligation to act with a view to preventing and organizing infrastructure to react adequately in crisis situations.
>
> (Cvrtila, 2009 in: Mikac, Cesarec, & Larkin, 2018: 28)

From a theoretical point of view, an important precondition for countering successfully the contemporary security challenges and risks is the adaptability of the early warning system, and risk management that is. Along these lines, the research showed that BiH, namely, its security structures, have very good capacities for the development of an efficient system for risk management incorporated by governmental and non-governmental institutions and organizations. Moreover, it proved that the security system of BiH has 22 organizational forms that have the elements for risk assessment and management (Lisica, 2011: 155–156). However, the research carried out as part of the study "Security risks and fundamental social values in Bosnia and Herzegovina" identified weaknesses and problems that eventually make it difficult to establish the system of the critical infrastructure protection on the path toward the establishment of a state register of risks. Therefore, the study emphasizes that problems are reflected in

> insufficient coordination, overlapping of jurisdictions and inadequately regulated exchange of data among the security system forces; IT disconnection and incompatibility of information systems and data bases; inadequate knowledge about risks, risk management, general and special standards and terminology of risk management.
>
> (ibid., 2011: 157)

For the purpose of a more complete picture, it is necessary to emphasize that the security structure of BiH is reflected through three levels, as follows: the State, the Entities, namely cantonal (Federation of Bosnia and Herzegovina) and local (Republic Srpska), and the Brčko District of Bosnia and Herzegovina. From the aspect of the critical infrastructure protection, the situation at Entities' and other levels is not at a satisfactory level, due to inadequate, non-integrated and non-systemic organization of sectors that may be part of critical infrastructure. The majority of administrative-territorial units in BiH started preparing documents on the Assessment of risk from natural and other disasters, and on strengthening the capacities of resilience to disasters. Through analysis we established that certain documents carry analysis of certain risks that may jeopardize critical infrastructure, however, there is no comprehensive analysis relating to the critical infrastructure protection. However, the Entity of the Republic Srpska regulated in 2019 by law the critical infrastructure protection by passing the *Law on the Security of Critical Infrastructures* (National Assembly of the Republic Srpska, 2019). That Law regards the following as critical infrastructure:

> systems, networks and facilities of special importance, which destruction or jeopardizing may cause serious disturbance to the free movement of people, transportation of goods and provision of services, it may have an adverse effect on the internal security, health and lives of people, property, the environment, external security, economic stability and continuous functioning of the authorities of the Republic.

Also, the same Law defined the sectors, within which critical infrastructure is designated (industry, energy industry, mining, information-communication infrastructure, traffic, healthcare, utility services, water management, finances, education, food and drink, production, storage and transport of dangerous goods, public services, cultural and natural assets) (ibid., p. 157). Still, the Law pertains solely to the territory of the Entity and left out the principle of possible connectivity of the components of the system with other sectors in Bosnia and Herzegovina. Therefore, it is possible to positively assess the passing of the Law dealing directly with the critical infrastructure protection, however, without adequate future cooperation and compatibility with other elements and levels in BiH, we cannot speak about the harmonized system that meets prerequisites for the functioning. Along these lines, the available study "Disasters Response Capacities Assessment and the Roadmap for Bosnia and Herzegovina" clearly identified the lack of cooperation, that is to say that "in a context of exposure to severe disasters that is characterized by rather limited resources the civil protection system also faces challenges related to the complex institutional governance architecture and systematic coordination between the various levels" (Italian Agency for Development Cooperation and United Nations Development Programme, 2018: 29).

Finally, although the need for and necessity to adopt minimum conditions for the critical infrastructure protection have been emphasized several times, the State of Bosnia and Herzegovina does not have an integrated and complete system of the critical infrastructure protection.

5.3. Public-Private Partnership

Public-private partnership in the security sector of modern countries is one of the key processes of building and protecting the critical infrastructure. Particularly so if one takes into account that the critical infrastructure is determined on the basis of sectoral and intersectoral criteria (Mikac, Cesarec, & Larkin, 2018: 27). When it comes to the area of public-private partnership in BiH in the function of the critical infrastructure protection, there is no uniform normative and legal framework. However, despite the non-existence of a uniform state framework, cooperation among potential critical infrastructure subjects and private security companies is present in BiH. Yet, the critical infrastructure managers continue to rely primarily on their own resources in all segments of protection (Cikotić et al., 2018).

Public-private partnership from the aspect of the critical infrastructure protection in energy and telecommunication sectors in BiH was established in the form of technical and physical protection. The energy sector in BiH for the production and distribution of electricity is composed of three, i.e. four providers owned by the State. In the territory of the Federation of BiH they are: "Elektroprivreda BiH" – Sarajevo and "Elektroprivreda HZHB" – Mostar, and "Elektroprivreda RS" – Banja Luka in the territory of the Entity of the RS. At the level of BiH, a state-owned enterprise "Elektroprenos BiH" was established and is solely in charge of the electricity transmission. Upon analyzing the work of these enterprises from the aspect of the critical infrastructure protection, it was established that they mostly rely on their own capacities in terms of technical and physical protection, i.e. civil protection. However, a possibility was left to engage private capacities in the event of natural or other disasters. On the other hand, a positive example of the public-private cooperation in the function of the critical infrastructure protection is the state-owned enterprise "Sarajevogas", which is in charge of gas distribution and has over 60,000 users, and more than 1,000 km of distribution network (Sarajevogas, 2020). The presented indicators clearly determine that this is the critical infrastructure in the area of the capital of BiH, Sarajevo. The mentioned subjects in the system of public-private partnership in the critical infrastructure protection use their own (public) resources (Civil Protection Staff, Service for Fire Protection and Workplace Safety, Service for the Physical and Technical Protection of People and Property) and private resources (construction companies, specialized legal persons for fire protection, private security agencies for the protection of people and property, specialized legal persons for support to the technical protection system). The process and level of inclusion of the private sector depends on the degree of endangerment of critical infrastructure. To summarize, all in all, the business entity "Sarajevogas" in all segments of protection has the support of a private partner (physical, technical protection, fire protection and protection from natural and other disasters) – public-private partnership. Also, the manager of the respective critical infrastructure has elaborate procedures for the operation in regular mode and in emergencies, for independent operation and for cooperation with the private protection segments and the public security segment (Smajić, 2017).

Nevertheless, the aggravating factor in achieving a public-private partnership and better cooperation in the function of the critical infrastructure protection stems from the lack of normative and legal regulations on the one hand, as well as from the dialogue among (private) partners on the other hand. Similar challenges in the implementation of a public-private partnership in the critical infrastructure protection show up in other states in the form of "vaguely established framework, distrust and week institutional power of the competent institutions" (Mikac, Cesarec, & Larkin, 2018: 32).

5.4. Cooperation with European Commission

The aspiration for membership in the European Union is the principal and fundamental political and social orientation of the State of Bosnia and Herzegovina. In that sense, in 2008 BiH had already entered into the first contractual relationship with the European Union by signing the Stabilization and Association Agreement (SAA). Upon entry into force of the SAA, in 2016 the State of Bosnia and Herzegovina applied for membership and currently has a status of a "potential candidate country" (European Parliament, 2020).

As to the current progress of BiH toward membership, based on the Progress Report, there has been no recorded progress in the area of the critical infrastructure protection. Also, in its Country Progress Reports the European Union did not specifically address or highlight the system of CIP or of civil protection in general. Further, it was noted that the European Union would provide continuous financial and other assistance for the implementation of the necessary reforms. However, BiH actively participates in international civil protection exercises (EU CAMPEX) and uses the EU pre-accession funds (IPA[2]; TAIEX) to strengthen the capacities in the area of civil protection, i.e. protection and rescue. Despite that, BiH still has no comprehensive strategic approach to building and protecting critical infrastructure.

5.5. Education and Training

Although BiH does not have the developed system of the critical infrastructure protection, over the recent years the public and academic sectors have increasingly identified the need for the development of a platform for the critical infrastructure protection. In that context, certain universities in BiH have already incorporated, or are about to do so, in their curricula the modules for the study of critical infrastructure and have implemented significant cooperation with the state institutions (Ministry of the Interior, civil protection administrations, private agencies and such like). Furthermore, the need for education on the critical infrastructure protection was recognized by private agencies too, which organize professional trainings, lectures on the need for the critical infrastructure protection. On the other hand, with the help of international organizations, such as OSCE Mission to BiH, BiH organized an exercise on the critical infrastructure protection from terrorist attacks in cyberspace in 2018, with the participation of experts from public institutions and private energy sector companies and agencies (OSCE in BiH, 2018). On the other hand, BiH has recognized the importance of cyber security and has developed certain curricula and teaching units at higher education institutions, which raise awareness about the importance of cyber security and contribute to the development of security culture in general. Furthermore, although there is no single state coordination plan for efficient cyber security management, still the existing curricula and courses at universities, as well as courses organized by international institutions, help in developing the necessary skills and know-how to manage cyber threats, all with the aim of creating more secure cyberspace. To that end, BiH is a host of the Southeast Europe Cyber Security Center (SEECSC), which offers training possibilities with the aim of enhancing the current capacities in the field of cyber security for the public and private sectors in BiH (Blueprint Energy Solutions GmbH, 2019: 52).

Finally, it is certain that BiH is just in its infancy when it comes to the development of the system of the critical infrastructure protection, including the area of education and training. However, the experiences of partner countries from the EU and NATO may certainly serve as a good example and help in building capacities and the ability to build an effective system of the critical infrastructure protection, particularly in the area of training and education of experts. Eventually it certainly "resulted in the identification of potential weak spots and the necessity to strengthen the identified shortcomings of certain procedures" (Mikac, Cesarec, & Larkin, 2018: 236).

Chapter Conclusion

BiH does not have the adopted *lex specialis*, a law that defines all critical infrastructure, instead the significance of this infrastructure is derived from certain laws and bylaws, according to individual branches of critical infrastructure, or infrastructure of vital relevance for the functioning of different levels of state organization and communities in BiH. Among the first measure to be undertaken, it is necessary to pass a law that defines the critical infrastructure in BiH. Furthermore, with the aim of establishing an efficient system of the critical infrastructure protection it is necessary to adopt a bylaw to define a uniform approach to risk analysis and make an optimum definition of the critical infrastructure protection. Also, it is necessary to prepare a state register of risk and connect all relevant institutions for the purpose of easier operative functioning in the event of disasters or such. On the other hand, it is necessary to design internal cooperation protocols in the critical infrastructure protection by different security segments holders within the security system in BiH (bearing in mind the peculiarity of the internal organization of BiH), as well as protocols on cross-border cooperation of BiH with the countries in the region and the world in the protection of own critical infrastructure. Finally, it is necessary to strengthen the inclusion of the private sector as a partner to public security structures, not only in the critical infrastructure protection, but of other infrastructures too.

Notes

1. A permanent solution to the establishment of peace and ending the war against Bosnia and Herzegovina was achieved by the General Framework Agreement for Peace in Bosnia and Herzegovina, initialled on 21 November 1995 in Dayton, and officially signed in the presence of witnesses in Paris on 14 December 1995. The Framework Peace Agreement, i.e. Annex 4 (the Constitution), provided that "the Republic of Bosnia and Herzegovina shall continue its state continuity in internationally recognized borders . . . consisting of two entities: the Federation of Bosnia and Herzegovina (51%), which comprises 10 Cantons, and Republika Srpska (49%)" (Smajić, 2018: 131–132).
2. Since 2019, Bosnia and Herzegovina as a partner country, in cooperation with the Republic of Croatia, is implementing the project "EU for Better Civil Protection – Capacity Building and Strengthening and Preparation of BiH for the EU Civil Protection Mechanism" (http://eu4bettercivilprotection.ba/en/).

Chapter 6

Serbia

Zoran Keković

Contents

This chapter provides an insight into the strategic, normative and institutional framework on CI protection in the Republic of Serbia. Before establishing the basic normative framework on CI in the Republic of Serbia in 2018, the identification of CI did not start from scratch, as some existing legal acts gave a solid starting point. As the country with EU perspective, before joining the European Union, the Republic of Serbia has an obligation to normatively arrange and regulate issues of identification, designation and protection of national and European CI. Therefore, the first step in the regulation of this field is the adoption of the Law on Critical Infrastructure, thus establishing the legal framework for definition, identification and protection of national and European CI.

6.1. Introduction

The concept of 'Critical Infrastructure' had appeared in Serbia, for the first time in 2011, in *the Regulation on the Content and Methodology for the Development of*

DOI: 10.4324/9781003273769-8

Protection and Rescue Plans in Emergency Situations (Government of Republic of Serbia, 2011). Article 8 of the Regulation highlighted the assessment of CI from the standpoint of natural disasters and other major accidents. However, neither this nor any other document gave a definition of the concept. Furthermore, *Guideline on Methodology for the Risk Assessment and the Protection and Rescue Plans in a State of Emergency* (Government of Republic of Serbia, 2012), established criteria for the assessment of 10 CI sectors with regard to their vulnerability to natural disasters and other accidents. Although the methodology contained the most comprehensive approach to the CIP in the national legislation, it was focused on identifying sources of threats and particularly on the consequences that a disturbance or interruption of the facility operation has on the economy and ecology. However, this methodology did not include an all-hazard approach, nor the measures for improving resilience that could reduce the adverse effects of natural and other disasters on the infrastructure, including the cascade effects caused by interdependencies.

In particular, the *Law on Defence* (National Assembly of Republic of Serbia, 2018a), with its related bylaws referred mainly to the defense industry of Serbia, but also to other industrial and infrastructure objects, which during the war, state of emergency or mobilization of the Serbian Army primarily provide the services and operations stipulated by the Ministry of Defence.

Other laws, bylaws and strategic documents relevant for the CIP are *the Law on disaster risk reduction and emergency management* (National Assembly of Republic of Serbia, 2018b), *National Strategy of Protection and Rescue in Emergency Situations* (National Assembly of Republic of Serbia, 2011), *National Security Strategy* (National Assembly of Republic of Serbia, 2019*), Law on information security* (National Assembly of Republic of Serbia, 2016a), *Law on Private Security* (National Assembly of Republic of Serbia, 2013), *Law on Environmental Protection* (National Assembly of Republic of Serbia, 2018c), *Data Secrecy Law* (National Assembly of Republic of Serbia, 2009), *Law on Planning and Construction* (National Assembly of Republic of Serbia, 2009), *Law on Water* (National Assembly of Republic of Serbia, 2018d), and other relevant documents.

As it can be seen, until the adoption of the Law on critical infrastructure, this issue has not been systematically regulated.

The adoption of *the Law on critical infrastructure* (National Assembly of Republic of Serbia, 2018e) is among the obligations of the Republic of Serbia in the process of EU accession. This law regulates national and European CI, identification and designation of CI of the Republic of Serbia, CI protection, jurisdiction and responsibility of authorities and organizations in the field of CI and information, reporting, decision making support, data protection, management and oversight of CI. Under this Law, national CI is defined as

> the systems, networks and objects or its parts whose disruption in operation or interruption in the delivery of the goods can have serious

> consequences for national security, health and lives of people, property, environment, security of citizens, economic stability, or jeopardize the functioning of the Republic of Serbia.
>
> (article 4)

According to the Law (article 5), the identification of CI shall be carried out on a sectoral basis in accordance with established criteria. While many developed countries identified over 10 CI sectors, the following sectors are identified as CI: energy, transport, water and food supply, public health, finance, telecommunications and IT, environmental protection, and public authorities functioning. In addition to the aforementioned sectors, CI may be designated in other sectors, at the proposal of the ministry responsible for a particular area, in accordance with this Law. It is interesting that Law stipulates that protection, storage, use, control and surveillance of CI under the jurisdiction of the Ministry of Defense and the Serbian Armed Forces is carried out in accordance with the *Law on Defense* and the *Law on the Serbian Armed Forces*. The Law also recognizes the OSP for risk management and liaison officers as important institutes in the CIP.

It is also emphasized that CI must take priority when developing planning documents in the field of spatial and urban planning, documents in the field of national security and in the area of risk reduction and emergancy managment (EM), especially in the part of preventive and emergency response activities.

In the following steps, it will be necessary to prioritize the identified CI sectors and regulate the aspects of the CIP that have shown to be particularly problematic in the European and global practice – public-private partnership (PPP) and exchange of classified information.

It has been necessary to define criteria for the identification of potential threats/hazards and interdependencies tailored to different CI sectors in line with international, European and national rules and standards. Various ministries, sectors and departments have different criteria and classifications of objects and facilities under their jurisdiction. The *Law on Defence* provides the definition of facilities that are of special importance for the national defense, while the Plan of Defence mentions hundreds of technical and technological systems, with the respective plans of defence. Therefore, it is possible that future CIP plans will be included in plans of defence.[1]

Another relevant law for the identification of CI is *the Law on Planning and Construction*, and its associated plans: *Spatial Plan of the RS*, followed by regional and local plans. Particularly important are the spatial plans of special-purpose areas, which almost completely coincide with CI.

Today, the Republic of Serbia faces the obligation and the need to pass a set of bylaws in order to more closely arrange the individual areas of the CIP. With this regard, *the Guidelines on methodology for the risk assessment of the natural and other disasters and emergency rescue and protection plans* (Government of Republic of

Serbia, 2019b), as another existing subordinated documents passed by *the Law on disaster risk reduction and emergency management* – already contains the rules for the risk assessment which includes facilities and other infrastructure of particular importance (CI).

6.2. Implementation

Preparation and adoption of appropriate legislation represents only the first step of the several necessary to ensure an effective and high-quality system of CIP. Besides the law on CI, a normative framework in accordance with the strategies and legislative on EM, defense, date secrecy, information and other security regulations provide a framework for many operational activities to meet the implementation of legislation and regulations in the area of CI.[2]

The Action Plan for Chapter 24 for the EU accession recognizes the Ministry of Internal Affairs (MIA) of the Republic of Serbia as the authority responsible for the Law on critical infrastructure. Within the MIA, the Sector for Emergency Management plays an important role as a centrally responsible authority both for the preparation and coordination of the legislative framework for CIP and its implementation. In any case, the most difficult task will be related to the harmonization of all other national strategic documents and their amendments to the relevant basis for the implementation of the area of CIP.

One of the most important legal innovations stipulated by the *Law on information security* is the establishment of the National Center for Security Risk Prevention (CERT), a body in charge of rapid response to incidents, as well as collecting and exchanging information on security risks of information and communication systems (Center for Euro-Atlantic Studies and OSCE Mission in Serbia, 2016).

The definition of the relevant criteria for determining CI in the Republic of Serbia is directly related to the extent to which systems, processes and activities will become CI. An excessive and unrealistic definition of the criteria for determining the CI could result in the creation of a theoretically defined system that in practice cannot be established. The struggle to increase the impact among the institutions of the state administration represents a very important risk, which does not have a negative impact only in the field of the criteria but also in the field of the authority of control and subsequent management of the system of CI, and, consequently, a higher impact and financial resources.

It is also necessary to take into account the relationships among businesses, especially those which are in competition on the same sub-sector of CI management.

The next moment is the relationship between the state and local level of governance which can result in a worse functioning of CIP. CI should be because of its importance in a systemic approach centrally managed. In this regard, CI resilience represents an interplay of national (as these are objects of national importance), community (they provide benefits and welfare to the communities that use their products and services) and organizational (as they are socio-technical systems) resilience (Keković & Ninković, 2020: 162).

Of course, it is necessary through various forms of cooperation to ensure that in developing a system of CIP the local community and their interests will also be taken into account in the Republic of Serbia. From this point of view, this coordination and a clear division of authorities and responsibilities is extremely important.

6.3. Public-Private Partnership

Public-private partnership (hereinafter – PPP) is among the key factors of the CIP process. The increase in the percentage of privately owned CI facilities is expected, taking into account global trends of market liberalization.

The *Law on Public-Private Partnership and Concessions* (National Assembly of Republic of Serbia, 2011, 2016b) does not recognize the term "critical infrastructure", however, it is clear that many facilities and services of public interest mentioned in this law will enter the framework of the Law on CI and future bylaws and regulations in this area. According to this Law, the PPP is long-term cooperation between a public and a private partner for the purposes of providing financing, construction, reconstruction, management or maintenance of infrastructure and other facilities of public interest and provision of services, of public interest, which may be contractual or institutional (article 7). The concept of 'concession' which regulates relations in various infrastructure sectors will surely be relevant for the PPPs in the field of CI.[3]

As one of the implementation challenges, it is difficult to predict that a framework or the basis for the effective functioning of PPP in Serbia will be at the appropriate level, which will enable more efficient cooperation. Significant attention will need to be devoted to the process of exchange of information and good practices and the establishment of appropriate channels of communications. Only the legislation for the transfer of key information is not a sufficient guarantor for an effective process. Confidence building is essential. The fact is that we have in this part at least two key categories of information, i.e., the information that is essentially important for ensuring national security and on the other hand, the information that represents important business data in the business environment, which may reduce the competitive advantage of a company which manages CI.

A serious shortcoming for PPP exists in the legal framework of information security. The law stipulates the establishment of an Information Security Coordination Body, composed of representatives of several state institutions, but not representatives of the private, academic or civil sector (Center for Euro-Atlantic Studies and OSCE Mission in Serbia, 2016).

When considering policy cooperation in the CIP in Serbia it should be stressed that the private security sector in Serbia is very active with private security companies which provide services in the field of CIP. At the same time, the necessary attention should be paid to the quality of services because it is essential that the private security sector reaches a certain level of competence for such operations.

Furthermore, private security companies are engaged in the CIP projects through the process of public procurement. It was noted that major problems arise in the process of public procurement of private security services, since the only criterion is the lowest available price, even though the EU Directive provides clear guidelines that mandatory criteria are the most economically advantageous tender, and the competitive dialogue.[4] This can have serious consequences for the CIP since many of CI facilities are protected by companies characterized by a low level of service quality, personnel training, equipment and others.

Although PPP is not the ideal model for all infrastructure projects, it is necessary to consider a joint action wherever possible and mutually justified. To this purpose, we propose the following: conclusion of long-term projects by the public sector, joint public and private funding, involvement of the private sector in the responsibilities of the public sector (procurement, construction, management, maintenance, etc.), creation of risk and responsibilities sharing models during the course of the partnership (Forrer et al., 2014: 4).

6.4. Cooperation with European Commission

According to the *Law on CI* (article 13), ECI can be identified in sectors designated by the European Commission. The designation of ECI on the territory of the Republic of Serbia is under the jurisdiction of the Government of the Republic of Serbia, at the proposal of the certain ministry, and on the requirement and with compliance of Member States. The government also informs interested Member States about the designation of ECI on the territory of the Republic of Serbia. If the CI of significance for the Republic of Serbia is placed on the territory of other Member State, government proposes to its responsible authority designation of the ECI.

Article 14 stipulates that ECI on the territory of the Republic of Serbia shall be protected in the same way as the CI of the Republic of Serbia, except where otherwise provided by European Union regulations. The previous article highlighted the importance of the determination of an appropriate contact point and, of course, the Serbian national body which will be responsible for coordinating the preparations for the protection of CI at the cross-border level. In this regard, most activities and cooperation have been undertaken by Department for Emergency Situation of the MIA.

Furthermore, many EU directives have been already implemented to different sectors of CI, by stipulating within related legislative framework areas harmonized with the EU.

In the framework of accession negotiations with Serbia, the European Union has so far mainly dealt with issues related to cybercrime, in chapter 24: Justice, Freedom and Security.

In the recommendations of the screening report of the European Commission related to the field of police cooperation and the fight against organized crime, the Commission identified the need to provide further specialized training and improve the capacity of law enforcement agencies in charge of fighting cybercrime. Through its recommendations, the Commission also recognized the need to establish close cooperation with the private and public sectors and academia (Center for Euro-Atlantic Studies and OSCE Mission in Serbia, 2016).

The Republic of Serbia actively participates in many EU projects, by using the EU pre-accession funds (IPA), EU civil protection mechanism and research initiatives to strengthen the capacities in the area of civil protection, i.e. protection and rescue, as well as on the municipality level where cities and local communities participate in different European Commission activities and mechanisms.

6.5. Education and Training

Human resources in each system represent one of the most important components of successful operation. In the case of the needs of personnel who will be able to successfully operate in the field of CIP, it is impossible to provide the personnel with the necessary multi-disciplinary knowledge in a very short period of time. Here the academic community has a very important role in the development of programs which on the one hand respect the regional and national-security environment, and on the other hand, the business environment, providing the managerial skills and experience necessary for CIP. This means that this segment will be the key in the measures of raising awareness and knowledge of two key groups that are formed by the strategic management in public administration and the strategic management in business organizations that manage CI.[5] In the case of certain multinational companies, which with their financial and other power press on the governments of the countries in transition, thereby achieving adoption of milder and more lax legislation, which they later can avoid. Unfortunately, the Republic of Serbia is not an exception in this process, which has very negative effects on the state of protection of CI in the country.

Scientific and research work on the RECIPE 2015 (project funded by the European Union to support further development of initiated system-building activities, with Serbia as a first beneficiary), has been resulted in Guidelines/Instructions for better and more efficient management and operational framework for CIP. The results of the RECIPE project should be taken into account when developing education and training curriculums and raising awareness of CI owners and operators and their strategic management in both the public and private sectors.

Chapter Conclusion

In the Republic of Serbia, the process of identification, prioritization, protection, resilience and legal regulation of the field of CI is at the very beginning. First of all, we refer to the adoption of the *Law on CI* adopted in 2018 that regulates this field. Bylaws pertaining to this law is in process of passing. In the process of establishing the normative framework, the lawmaker most frequently takes over the *Acquis Communautaire* (in the EU context) and opens a public discussion with all stakeholders.

After the CI-related law and bylaws are adopted and the CI sectors and facilities identified, the following step will be the prioritization, as not all CI sectors and facilities are equally critical from the aspect of the disruption of their operations or interruption of supplies of goods and services. Certainly, the existing problems and challenges will not be resolved with passing of this Law. The CI sectors and facilities identification and prioritization, adoption of methodologies for the CI risk assessment, PPP in the field of CIP, and exchange of classified data are the main, but not the only challenges that lawmakers, CI owners and operators and other stakeholders in Serbia will face in the coming period. These challenges may be overcome by passing the bylaws, adoption of amendments and harmonization of other relevant laws, by education and raising awareness of CI owners and operators, by adoption and implementation of national and international standards and improved cooperation with academic institutions. It is extremely important since the process of prioritization has caused much disagreement between CI owners and operators on one side, and lawmakers on another.

The establishment of the PPP mechanism in the CI protection and resilience arena is of particular importance, as during the process of liberalization more CI facilities will pass into private ownership, while the private security sector plays an increasingly important role.

Furthermore, the question of classified data sharing in the CIP system is a complex one and will need to be considered carefully by all stakeholders. An efficient and secure system of sensitive and classified data exchange needs to be established, with the accent on the 'horizontal' approach (exchange of data between sectors and CI systems), in comparison with the 'vertical' one, which is still prevalent but is not quick and efficient enough.

After it is clearly defined, what we understand under CI protection and resilience, and what it needs to achieve, it will be necessary to devise a strategy for its implementation and to provide the political will to implement its objectives.

Notes

1. In addition to the Law on Defence and Plans of Defence, the following bylaws are also relevant for future identification and classification of CI in Serbia: Decision on Facilities of Particular Importance for National Defence (Government of Republic

of Serbia, 2008); Decision on Identification of Large Technical Systems Important for National Defence (Government of Republic of Serbia, 2015); Decision on Identification of Products and Services of Special Importance for the National Defence of the Republic of Serbia (Government of Republic of Serbia, 2008).
2. Some of them: training for security personal in critical infrastructure facilities, conducting risk assessments, education process and programs at universities related to critical infrastructure protection, provide good investment and foundation for the implementing of critical infrastructure policy in Serbia.
3. According to the law, the state PPP Commission, as the most important state body for the approval of PPP projects, is composed of representatives of various ministries, under whose competencies are areas that will belong to future CI sectors – such as the Ministries of energy, transport, mining, etc., their representatives will be directly involved in projects related to CI belonging to their competencies.
4. As the price of private security services in Serbia is the lowest in Europe, the quality of services is questionable if the cheapest offer is chosen.
5. Unfortunately, the situation in transition countries, including the Republic of Serbia, especially when it comes to companies that are in majority owned by the state, primarily focuses on satisfying narrow political and economic goals, which often are not based on good governance and concern for the continuous operation of critical infrastructure.

Chapter 7

Montenegro

Olivera Injac

Contents

This chapter provides an insight into the strategic and normative framework for CIP of Montenegro and analyzes a set of documents that provide specific norms, directives, jurisdictions and responsibilities. It provides insight in tasks for implementation of national CIP governance system, based on *Law on determination and protection of critical infrastructure of Montenegro*, which determined responsibilities for CI stakeholders. Capacities for public-private partnership in the field of CIP are presented through the cases of two public companies for energy ("Elektroprivreda of Montenegro", two-thirds state-owned company for production and supply of electricity) and transport ("Monteput", state-owned company for transport management and monitoring). Further, this chapter includes cooperation with the European Commission, insight into the capacity for competency building and conclusion.

DOI: 10.4324/9781003273769-9

7.1. Introduction

Montenegro has been an EU candidate state since June 2012 when the European Council endorsed the decision to open accession negotiations. For more than nine years now the country has been going through the negotiation process. The national strategic orientation for critical infrastructure protection (CIP) is articulated within the document *Montenegro National Security Strategy* (NSS) published in 2018, which prioritizes the development of an efficient CIP system and the strengthening of resilience through the cooperation of different actors (state institutions, civilian and private sector) (Government of Montenegro, 2018a: 12). The national *Law on determination and protection of critical infrastructure of Montenegro* was adopted in December 2019. The adoption of the Law on CIP, and *Law on crisis management*, is a continuation of the activities for resilience-building which were planned and outlined in NSS (ibid., 2018a: 53).

Even though Montenegro has not addressed CI in the normative framework until now, in the last decade the state has worked to establish a set of legal and strategic documents which refer to the areas of CIP, such as:

- *Law on Protection and Rescue*
- *Law on Protection of Persons and Objects*
- *Law on State Property*
- *Law on Cyber Security*
- *National Security Strategy*
- *Strategy on Cyber Security (2018–2021)*
- *Strategy for Disaster Risk Reduction (2018–2023)*
- *Law on Determination and Protection of Critical Infrastructure of Montenegro*

Also, for critical infrastructure in Internet Technology (IT) and cyber security, *Methodology on the critical infrastructure selection* (2014) is an important document.

The *Law on protection and rescue* was first adopted in 2007, as a general legal act for natural disasters, technical, technological and other accidents, which was further advanced with amendments, a few times since. It regulates prevention efforts, preparedness, professional protection, and rescue of civilians, material goods and cultural assets. This Law outlines and defines priorities for protection – health and human lives, material goods, cultural heritage, and environment. Also, indirectly, this law manages CIP, with regulations that public (national and local) and private institutions are responsible for insofar as preparing risk assessment and outlining and arranging plans for protection and rescue (Government of Montenegro, 2016a: article 33–35, 47–49).

Law on state property, adopted in 2009, defines natural resources, state and public goods, and infrastructure that are of interest for state protection (Government of Montenegro, 2009: article 11–19).

Law on protection of persons and objects, adopted in 2018, is the legal framework for the mandatory technical and physical protections and security that private companies must provide for their facilities whose damage could cause serious risks for citizens (Government of Montenegro, 2018b: article 13). Some of the mandatory protected objects, which belong to CI, are facilities for oil and gas storage, water supply, production and storage of food, production and distribution of electric energy, production and storage of radioactive and other harmful substances, transportation infrastructure (bus and train stations, airports, etc.), financial institutions, post and telecommunications, and others.

The *Law on cyber security*, adopted in 2016, defines IT CI in the public sector and addresses risks and responsibilities (Government of Montenegro, 2016b: article 14a). Critical IT infrastructure is defined as "information systems whose disruption or destruction could jeopardize life, health, safety of citizens and state functioning or from whose functioning depends public activities" (Government of Montenegro, 2017a: 14).

The NSS, adopted in December 2018, has recognized that attacks on CI could cause significant effects on national security, especially in the case of hybrid attacks (terrorism, propaganda, disinformation, cyber-attacks). The Strategy indicates ICS and Information and Communications Technology (ICT) infrastructure out of the territory of the country as the most risky areas but also the necessity of resilience and CIP in the area of energy security in particular (Government of Montenegro, 2018a: 17–20). The most important statement in the Strategy refers to the interdependencies of the cyber domain with several CI sectors (transport, telecommunications, health and social system, media, financial system, energy, etc.).

The *Strategy on cyber security (2018–2021)* has aimed to strengthen capacities for the IT CI protection, and the general security of critical infrastructures, in addition to achieving national cyber security goals. Eight sectors of IT critical infrastructure are addressed in the Strategy, with one or more subsectors in which ministries or government agencies have control of oversight insofar as the responsibility for their regulation (Government of Montenegro, 2017a: 27). Considering the fact that the biggest part of the critical IT infrastructure is held and managed by the private sector, the partnership between the public and private sectors is a priority (ibid., 2017a: 16).

The *Strategy for disaster risk reduction (2018–2023)* aligns critical infrastructure in the area of protection and risk reduction from natural and other disasters and highlights measures for protection and prevention. Critical infrastructure is referenced in the context of the risks to public health, which is a key priority of the Sendai Framework,[1] with a focus on reducing the mortality rate and the well-being and safety of the health facilities and hospitals (Government of Montenegro, 2017b: 97).

The *Law on determination and protection of critical infrastructure of Montenegro* is significant in relation to the building of the national system for critical infrastructure protection. The initial activities on CIP regulations started in 2019, although the intention existed before because previously mentioned legal documents do not directly regulate this field. This law was adopted on December 26, 2019, and it not only prescribed identification and definition of the critical infrastructure but also provided guidance for jurisdictions and responsibilities.

7.2. Implementation

Until now, the legal system of Montenegro has not specifically regulated critical infrastructure. National governance for the CIP system is expected to be developed based on the legal framework of the *Law on determination and protection of critical infrastructure of Montenegro*, which was adopted in December 2019. The Ministry of Interior is responsible for the law's implementation, and the next important step is that Law is complemented by secondary legislation identifying the list of infrastructures considered as critical for state security.

The aim of the Law is to identify, determine and protect critical infrastructure, as well as to define competencies, responsibilities, and other issues of importance for the national CIP system. Critical infrastructure is defined as

> systems, networks, facilities, or their parts, located within the territory of Montenegro, whose interruption of functioning, or interruption of delivery of goods or services through these systems, networks, facilities, or their parts, could cause serious consequences for the national security, human health and life, property, environment, safety of citizens, economic stability, but also activities of public interest.
>
> (Government of Montenegro, 2019a: article 2)

The Law recognized main sectors of critical infrastructure: 1. energy, 2. transport, 3. water supply, 4. health, 5. finance, 6. electronic communications, 7. information and communications technology, 8. environment protection, 9. functioning of the state institutions and 10. other areas of public interest (Government of Montenegro, 2019a: article 6).

Although a system for the critical infrastructure protection of Montenegro is not managed yet, a system of state institutions could be an appropriate framework for the CIP. Hence, this chapter will present legal norms and expected measures for the Law implementation.

National critical infrastructure encompasses multiple sectors which are interdependent. Determination of the criteria for critical infrastructure within the sectors is the responsibility of the ministries that are in charge of specific sectors (ibid.: article 11).

The decision on the criteria for determination within the sectors of critical infrastructure will be made by the government, based on the proposal of the MoI, which will coordinate activities of responsible ministries on CI determination. This could create obstacles due to possible non-synchronized approaches and the length of the process.

All CI operators are obliged to appoint a "coordinator for critical infrastructure" (ibid.: article 18), an employee who has an official certificate for the CIP, and qualifications for that position.

The operator is responsible for the preparation of a security plan, which is the most significant for the CIP because it should contain not only risk analysis and risk management but also measures for reducing the threats (ibid.: article 14). The MoI is responsible for the plan approval.

Decision-making in an emergency situation involving serious risks, disruption or destruction of critical infrastructure is the responsibility of the "coordination team for CIP" (ibid.: article 21); the national CIP structure, in general, is defined by the Law (coordination team, operators and coordinators for CIP) (ibid.).

One of the Law's chapters focuses specifically on the critical infrastructure of the European Union, considering the fact that Montenegro is a candidate state in the process of negotiation (ibid.: chapter 4: article 24–29).

The *Methodology for selection of the critical information infrastructure* was prepared in 2014 and consists of an evaluation of sectors and activities for implementation of the CIP system. This Methodology is the initial step for the identification of national critical IT infrastructure. Therefore, the identified sectors of critical IT infrastructure and responsible ministries, based on dependence on ICT, are as follows: 1. Telecommunications and IT (Ministry for Information Society and Telecommunications[2]), 2. Banking and finance (Ministry of Finance), 3. Energetics (Ministry of Economy), 4. Health care (Ministry of Health), 5. Agriculture, food safety, forestry and water management (Ministry of Agriculture and Rural Development), 6. Security and defense (Ministry of Defence, MoI, National Security Agency, Ministry of Justice), 7. Transport (Ministry of Transport), 8. State institutions and the government[3] (Ministry of Information Society and Telecommunications of Montenegro, 2014: 11–12).

Considering the fact that critical infrastructure includes broad areas and sectors, recommendations from the Methodology include categorization and the preparation of a detailed analysis, since IT critical infrastructure is divided into certain subsectors and different operators (both public and private). In 2017, the analysis on the identification and interdependence of IT critical infrastructure in public and private sectors was completed.

7.3. Public-Private Partnership

Since critical infrastructure protection is a "broad and dynamic activity", stakeholders use public security institutions or private security companies for CIP (Mitrevska, Mileski, & Mikac, 2019: 29).

Public-private partnerships in the field of CIP are established in many sectors in Montenegro, most fundamentally in the form of the technical and physical protection of critical infrastructure. Bearing in mind that companies in the energy and transport sectors are still state-owned in Montenegro, within these two sectors certain forms of public-private partnership in the field of critical infrastructure protection are present. In this chapter, we will be presented with two examples of state-owned companies from the energy and transport sectors:

- "Elektroprivreda of Montenegro", two-thirds state-owned company for production and supply of electricity
- "Monteput", state-owned company for transport management and monitoring

"Elektroprivreda of Montenegro" (EPCG) is a national electro-energetic company, which deals with the production and supply of electricity. Montenegro has over two-thirds or 70.16% ownership. Some of the stated strategic goals of the company are efficient and reliable production, electricity supply, efficiency, trade, resource management, environmental protection, protection and safety of employees at work, and more. EPCG has a complex electro-energetic system and therefore applies mechanisms and instruments for its physical, technical, information and organizational protection. As a large system, EPCG generates and exchanges a significant amount of electronic and other information required for decision-making and functioning. Therefore, protection measures and adequate policy of information security is of significant importance for the functioning of the electro-energetic system of Montenegro. The company has its own facility security service, and – according to the *Law on protection of persons and objects* – their facilities for production, transmission and distribution of electricity are deemed as national critical infrastructure, which means that they must have physical and technical protections in place.

In 2016, EPCG hired a consortium of private companies to estimate the need for critical infrastructure protection and develop plans for the physical and technical protection of their facilities. Therefore, it is estimated that EPCG is a very demanding system that requires full protection of their complexes of power plants two hydro "Piva" and "Perućica" and one thermal "Pljevlja", as well as administrative buildings for distribution and supply, Directorate building, but also and 89 substations 35/10 KW (EPCG, 2016: 14). Also, it considers that EPCG improves technical protection of a control system for access into the administrative buildings in order to protect facilities from unauthorized access. The EPCG engages private IT companies for that sort of protection (EPCG, 2016: 14).

"Monteput" is a state-owned company with a broad spectrum of jurisdictions, but the two biggest priorities of the company are the following: the management of the "Sozina" tunnel (construction, electrical, IT and mechanical maintenance, transport management, toll, etc.), which connects the continental and southern parts of the country, and the management of the biggest state investment, the construction of highway "Bar-Boljare", whose contractor is the Chinese company China Road and Bridge Corporation. The Monteput company has contracts for technical and physical protection of critical infrastructure with private enterprises, mainly for the protection of the critical objects and their video surveillance.

The public-private partnership in the CIP area is currently established just in the field of technical and physical protection of the critical infrastructure objects, and by considering these two cases, it is necessary to expand areas and capacities for partnership.

7.4. Cooperation with European Commission

The Law on the determination and protection of critical infrastructure of Montenegro fits with the *Council Directive 2008/114/EC* on identification and designation of ECI and progress of the system protection. The Law chapter on the EU critical infrastructure will be implemented at such time as Montenegro becomes an EU member (Government of Montenegro, 2019a: article 38).

Montenegro has been in negotiations with the EU since 2012, and during that process, many national legislative directives have been implemented within many different sectors of critical infrastructure. For instance, within the chapters, transport policy, energy, environment, consumer and health protection, and others, are formalized legal acts and established specific areas of cooperation with the EU. Through the process of negotiation with the European Commission, active cooperation exists in some of the sectors, such as within the sub-committee on transport, energy, environment and regional policy (Government of Montenegro, 2019b). This sub-committee has the greatest contribution through the regular meetings, and since the beginning of the negotiation process, Montenegro has made significant progress in the areas covered by this sub-committee, as demonstrated by opening and initialization of activities in the following chapters: 14 Transport policy, 15 Energy, 21 Trans-European Networks, 22 Regional Policy and Coordination of Structural Instruments and 27 Environment. The aforementioned are recognized and valued as significant contribution to the improvement of the protection of national critical infrastructure.

Up until now, the greatest cooperation was promoted by the Directorate for Emergency Management of the Ministry of Interior, which participates in different European Commission activities and mechanisms. A significant instrument for cooperation is the EU Civil Protection Mechanism, in which Montenegro has participated since 2014, with the task to improve prevention, preparedness and response for disasters. A part of that is also in cooperation with the European Commission Directorate-General for Humanitarian Aid.

In the national *Disaster Risk Reduction Strategy 2018–2023*, important parts of the Sendai Framework are highlighted, which are climate changes, environment and natural resources, and economic management and critical infrastructure in the health sector (Government of Montenegro, 2017b: 55–56). In addition to achieving efficient implementation of the Sendai Framework, Montenegro has cooperation through the *European Forum on Disaster Risk Reduction* (EFDRR),[4] in the field of sharing information, risk reduction, assistance, and so on. The intent of EFDRR is to enhance disaster risk prevention and European ability for risks and vulnerability reduction.

7.5. Education and Training

A national system of the CIP cannot be established without adequate knowledge and competencies (Mikac, Cesarec, & Larkin, 2018: 113), which means that programs at the academic level should provide theoretical and professional guidelines. Informal learning (training, conferences, etc.) should contribute to improving skills and readiness.

At the academic level, Montenegro has the capacity and capability to educate on critical infrastructure. For instance, the University of Donja Gorica (Podgorica) is an academic institution with modern programs of study, and it has the potential to be the central point for education and research in the field of CIP. A few programs offer academic education which indirectly deal with the issues of protection of critical infrastructures, such as Faculty of International Economics, Finance and Business, Faculty of Information Systems and Technologies, Faculty of Polytechnics, Faculty of Food Technology, Food Safety and Ecology, and Faculty of Humanistic Studies – Department of Security Studies.

The only program for Security Studies in Montenegro exists at the University of Donja Gorica (Podgorica). Academic disciplines and study areas that indirectly teach on CIP at the bachelor level include national security, cyber security, security culture, crisis management and international terrorism. Also, at the postgraduate level is a program for cyber security, which directly teaches students on IT critical infrastructure.

Informal education programs exist at the Urban Protection Training Center, Montenegro (UPTC), which train persons who intend to work specifically on protection tasks, in accordance with the *Law on protection of persons and objects*. While UPTC primarily trains those entering the protector profession of person and property, it also organizes training for different forms of protection (technical protection, fire protection, and safety at work) and other sorts of informal adult training and exercises (UPTC, 2019).

In the last few years, the Ministry of Public Administration has organized intensive training activities for IT sector protection.[5] Most of the organized training and conferences have been on the topic of cyber security and data protection.

Certainly, not a lot of training activities on CIP have been realized so far. Accordingly, one of the national priorities for CIP capacity-building should be the organization of conferences, seminars and other activities, in addition to improved knowledge and skills of decision-makers, managers and personnel. Also, at state level, the system for formal and informal training programs in this field is not built yet. Hence, it is necessary that national trainings for CI coordinators are developed soon, because these roles are often the first in line in the responsibility chain for national critical infrastructure protection.

Chapter Conclusion

Montenegro should intensify activities to overcome any delays in the law adoption and for the establishment of the CIP system. In terms of the lack of national capacities, Montenegro, as NATO member from June 2017 and candidate state for European Union from 2012, has experienced significant delay in recognizing and protecting critical infrastructure, and for that reason, the state, in the short term, should establish certain national procedures and systems in the following order of priority: 1. identification of national critical infrastructure, 2. development of criteria for critical infrastructure by sectors, 3. building the structure (coordinators, security plans, etc.), 4. training of personnel engaged in critical infrastructure systems, 5. training of CIP system within the sector and interdependently, which can be able to respond to protect or recover tasks in case of crisis or emergency. The important future activities of the country in this regard should be to determine basic criteria for identification, because it facilitates the entire process of the critical infrastructure determination.

Determination of the critical infrastructure and capacity-building for CIP are two of the highest priority activities, which are expected to be finished by Montenegro shortly after the adoption of the *Law on determination and protection of critical infrastructure*. Successful Law implementation is significant at this stage because weaknesses will be uncovered during this process and, certainly, it with create a path, moving forward, for building efficient protection systems.

Notes

1. The Sendai Framework for Disaster Risk Reduction, 2015–2030 (Sendai Framework), is a UN Disaster Risk Reduction Framework 2015–2030, which is adopted in March 2015 in Sendai (Japan).
2. In November 2016, by the government decision, the Ministry of Public Administration obtained jurisdiction of the Ministry for Information Society and Telecommunications and continues to coordinate future activities on the analysis on critical IT infrastructure.

3. Ibid.
4. The EFDRR should enable Sendai Framework implementation and consists of the National Disaster Risk Reduction Focal Points and representatives of National Platforms in the European region, UNDRR Europe Office, Council of Europe (EUR-OPA), European Commission, and sub-regional organizations/institutions as agreed by the forum.
5. Previously organized by the Ministry for Information Society and Telecommunications.

Chapter 8

North Macedonia

Marina Mitrevska

Contents

While some countries such as Croatia, Slovenia, Sweden, the Netherlands, Serbia and others advance in the development of national policies for critical infrastructure protection, the Republic of North Macedonia is still seeking its place and role in this area. Among the institutions of the Government of the Republic of North Macedonia, the most active in this regard are the Ministries of the Interior and Defence, which strive to promote the importance of this topic and direct their efforts toward the development of the area by strengthening resilience and critical infrastructure protection.

This chapter provides an insight into certain laws and bylaws and the ways in which they regulate the field of critical infrastructure protection in North Macedonia. The analysis indicates that these acts are a good basis for initiating a normative framework as a fundamental basis for the identification and designation of critical infrastructure. Furthermore, they are a good basis for specifying future activities for creating an optimal system for critical infrastructure protection in North Macedonia.

DOI: 10.4324/9781003273769-10

8.1. Introduction

European Union is one of the most attractive economic, political and international phenomena and a fundamental element of the security architecture in the European area. Being established as a peace project, today it has built a specific approach and diverse capacities to tackle the challenges of the 21st century. The experience so far shows that one of the challenges for the European Union is its approach to the protection of critical infrastructure. That is to say, it is seeking its place and role in this area, striving to promote the importance of this issue, to ensure cooperation among the Member States, as well as to accelerate the exchange of knowledge and experience and guide Member States in their efforts to develop critical infrastructure framework. This approach should also be primary for the candidate countries for membership in the European Union, as in the case of the Republic of North Macedonia. Notwithstanding, this should be the first impetus since the challenges at the level of the Republic of North Macedonia regarding this area are multilayered and the solution is slow. The necessity to create own identity in this area is clear, and of course, it will be useful to take into account the positive experiences of some Member States of the European Union and the European Commission.

Given in a specific time frame, this chapter presents an analysis of the current situation in North Macedonia related to the initiative to develop a critical infrastructure protection system. In practice, the need to establish a normative framework as the main basis for identification has become increasingly relevant and a priority topic on the security agendas of government institutions, the scientific and expert public. Regarding the situation in North Macedonia, there is a fundamental position that the overall protection of the state in terms of preserving the functioning of critical infrastructure must be based on the "protection package" of the infrastructure. In order to do so, one requires an analysis of the open issues and challenges that North Macedonia faces. The analysis should lead to normative solutions initiative, aligned with other sector strategies, laws, programs and plans. Therefore, North Macedonia is undertaking and implementing a large number of strategic and normative activities that emphasize the importance of critical infrastructure and the need for adequate protection. Among the most important activities, the following are being listed: defining objects as critical infrastructure, defining measures for their protection and safety, and defining duties and responsibilities. It is particularly important to note that the designation of critical infrastructure in North Macedonia is not in line with the framework for normative regulation of issues related to identification, designation and protection of the ECI. This especially refers to the obligations arising from the EU *Council Directive 2008/114/EC* on the identification and designation of ECI and the assessment of the need to improve their protection. Many analysts agree that there is a lack of clear specifications of the critical infrastructure term. In that sense, it is generally accepted that in the specification of objects as critical infrastructure, one should start by analyzing the *Decision on designating persons and objects for protection*. This Decision was adopted based on the *Law on Internal Affairs*. The Decision precisely lists the objects of interest for the security of the Republic of Macedonia, such as electric power, postal and shipping, railways, airports, and water supply, and the *Decision on designating the legal entities that are obliged to private security*.[1] Practice shows that in order to be able to protect the critical infrastructure in North Macedonia operatively, professionally and efficiently, Section IV of this Decision prescribes mandatory private security for legal entities, especially when it is in the interest of attaining security of the state (Mitrevska & Mikac, 2017: 49). In particular, several sectors have been defined, namely, energy (production, transmission and distribution of energy), water supply, environment, Macedonian Radio and Television, electronic and print media, National Bank of the Republic of North Macedonia, and other legal entities registered for performing banking activities.[2]

The Republic of North Macedonia has no formally defined list of critical infrastructure, it is not regulated by legislation and there is no identification and protection of critical infrastructure. Hence, an established precise EU specification of critical infrastructures and the stated solutions in the EU and NATO Member States, especially the examples of Slovenia and Croatia, would be of great benefit to future activities when establishing a formal framework for the protection of the national critical infrastructure. The analysis shows that the protection and security of critical infrastructure in North Macedonia should be directed toward several key sectors, such as the energy sector, information technologies, water systems and air traffic. The energy sector in North Macedonia is regulated in accordance with the *Law on Energy*. Here, as a priority, we would highlight strategically the most important companies, such as "ELEM" (Macedonian Electric Power Plants) and "JSC MEPSO" (Macedonian Electricity Transmission System Operator), which together with their capacities represent the backbone of the energy system. In the oil industry, "JSC OKTA", Makpetrol and Lukoil have priority in the protection because they play a significant role in the sale, supply and distribution of petroleum products in North Macedonia. In the IT sector, we would single out strategically the most important critical infrastructure, that is, "Makedonski Telekom" and "A1". These companies with their full capabilities represent the backbone of the landline and mobile network and the most sophisticated information technologies. Water systems in North Macedonia are regulated in accordance with the *Law on Waters*, which emphasizes the wide range of measures for security and protection of surface waters, lakes, reservoirs and springs, water management facilities and so on. The Law covers the regulation of the availability of sufficient quantities of safe and clean drinking water, the supply of safe drinking water, prohibition or restriction of use in case of its contamination, and taking measures to continuously ensure the quality of drinking water. Air traffic in North Macedonia is regulated in accordance with the *Aviation Law*, according to which the organizations involved in the safety of civil aviation at the national level are the Civil Aviation Agency, the Ministry of the Interior, and airport operators and air carriers.[3] These three prerequisites are particularly important in terms of effective security of this critical infrastructure: continuous development, implementation of legal regulations, and continuous undertaking of measures, programs and procedures. With a view to achieving a standardized level of aviation safety, a comprehensive policy supported by legislation should be adopted by the body responsible for security (usually through the Civil Aviation Agency), which will be implemented by all entities involved in any civil aviation security structure. Each of the aforementioned entities, police services, air carriers, intelligence services, and so on, must have clearly defined policies, procedures, standards of practice and methods of application in accordance with the guidelines of the state. The process should begin with the adoption of a proposal for establishing a National Security Committee and a Committee for Airport Safety and it is important to emphasize the need for establishing new efficient bodies to implement the policy and standards for implementation of security measures in a coordinated way[4] (Alcheski, 2016: 213–214).

When undertaking and implementing strategic and normative activities, which would emphasize the importance of critical infrastructure and the need for its adequate protection, most importantly is the obligations that North Macedonia should undertake not to be mere "ad hoc" mechanisms. On the contrary, they should represent a comprehensive reflection of the European approach to critical infrastructure protection. Therefore, North Macedonia must have capabilities to offer an explicit alternative to "critical infrastructure threats".

8.2. Implementation

Regarding critical infrastructure protection, North Macedonia has paid attention to the issue in the last period in its strategic documents and normative acts in the field of defense and security. These acts represent a prerequisite for forming a complete framework of legislation and procedures in order to begin the process of developing a critical infrastructure protection system. Some of those strategic documents are the *National Cyber Security Strategy of the Republic of Macedonia* (2018–2022), *Defence Law* (in Parliamentary Procedure), *Law on Internal Affairs, Crisis Management Law, Protection and Rescue Law* and *Law on Private Security.*

The initiative to develop the entire process of protecting critical infrastructure systems and identifying the need for its establishment is encouraged by the preparation and adoption of the *National Cyber Security Strategy.* In this document, the critical infrastructure area is perceived in terms of protection against cyber incidents and abuses, making these threats one of the most serious ones to critical infrastructure. The Strategy devotes much space to the critical communications and information infrastructure brought about by cyber crisis management. It also emphasizes the need to strengthen national capacities for cyber security prevention and protection and implement activities to raise national cyber security awareness.

This Strategy frames the critical infrastructure in a wider range, stating that disruption to the provision of certain ICT-dependent services may be critical to the functioning of the state. The Strategy defines cyber-physical threats to critical infrastructures, such as increased number of cyber-attacks, including industrial cyber espionage, cyber vandalism and vulnerability identification in the energy sector, transport systems, and other parts of the CII.

Moreover, several different normative legal solutions are important for analysis. First, the *Law on Internal Affairs*, which regulates the obligation of the police to protect important objects that are specific, i.e., part of critical infrastructure. In the second case, it is the *Defense Law*, which has established a process of identification and protection of objects critical to the country's defense. Certain sections and views in the draft law on defense deserve serious analysis. These refer to the views that define the rules and criteria for designating and protecting objects of particular importance to the defense of the state in the context of the *Defense Law* with a clearly defined methodology for risk assessment and analysis and a plan for the protection of military and other designated objects (art. 42 and 43). The analysis indicates that the majority of objects will overlap with potential critical infrastructure objects primarily due to their exceptional similarity and to a lesser degree their equivalence as well. In other words, it will raise a number of questions with different dimensions. First, the objects need to be included and delineated in the new critical infrastructure law. Second, two parallel processes emerge for objects to be equally important, so as to be classified as objects of special importance to the Ministry of Defense of the State, that is, critical infrastructure objects, with the requirement of two different coordination systems.

The third case is the *Law on Private Security*, which, in its normative framework, prescribes the government to determine which legal entities are obliged to private security. Namely, when their activity is related to handling radioactive substances, other dangerous substances for people and the environment, objects and facilities of particular cultural and historical importance and in other cases when it is in the interest of the security, i.e., defense of the Republic of Macedonia (art. 44). This Law also initiated the need to adopt a *Decision on designating the legal entities obliged to private security*. This Decision enables private security agencies with their capacities for the protection of capabilities, both human resources and technical solutions, to contribute to ensuring a high level of systemic critical infrastructure security, and above all, to contribute to preventive measures. Therefore, the Decision[5] specifically lists the security of legal entities, whose activity is related to handling radioactive substances or other harmful substances to humans and the environment, manufacture and wholesale of medicines and medical devices, production and trade of flammable liquids and gases, transportation of dangerous substances and handling objects, and facilities of particular cultural and historical importance.[6]

Another law that deserves attention is the *Law on Crisis Management*. This Law developed policies that involved the main actors: the government, state administration bodies and state authorities, the army and the protection and rescue forces.

Although this Law does not include the area of critical infrastructure, it is actually strongly recognized and underlined that "material goods" represent the infrastructure (Mitrevska, Mileski, & Mikac, 2019: 164). Given the fact that it also refers to the levels of competence of the Ministers of the Interior, Health, Defense, Transport and Communications, and Foreign Affairs, then, it is clear that the government has the leading role in critical infrastructure protection as well. The Law devotes much space to the responsibility of each of the institutions involved in the crisis management system, in terms of measures and activities for collecting information and identifying security risks, including those that endanger critical infrastructure security. In terms of legislation and powers, it is important to emphasize that based on their assessments the institutions encompassed in the crisis management system precisely determine the goals, tasks and implementation of the necessary actions for prevention, early warning and crisis management. It should also be noted that in the performance of their duties prescribed by Law, the actors in the crisis management system are obliged to communicate and coordinate among themselves and cooperate with the Crisis Management Center (art. 37). Furthermore, the Crisis Management Center has fully complied with this legal provision by establishing a web-based integrated GIS. It comprises an inventory of critical infrastructure elements that is consistent with the *Regulation on Methodology for Making Security Assessment of the Republic of Macedonia Against All Risks and Dangers* (adopted in 2005) and *Methodology for assessment of damages from natural and other disasters* (adopted in 2001). The importance of this Law can be seen in the fact that it is the basis for planned, timely, comprehensive and coordinated decision-making, guidelines and recommendations for taking preventive measures, as well as for the most optimal dealing with a crisis. Based on these, an assessment is made of the security threats to North Macedonia including all risks and dangers (adopted by the Government of the Republic of North Macedonia on 05.11.2019). Within the document, *Security Threats Assessment of the Republic of North Macedonia* including all risks and dangers, the section on "risk elements" comprises critical infrastructure in the broader context: hydrographic network, water management infrastructure, communication network, traffic and communications, road network – by types of roads and other infrastructure. This document provides a part on vulnerability assessment and exposure to specific risk elements, by identifying the typology of objects and their number, infrastructure and critical objects sensitive to tangible hazards, by mapping, if possible (Section 2.2.), the impact assessment is stated precisely (vulnerability and exposure) from specific hazards on objects and infrastructure (Section 2.4.).

The next important document is related to the importance of protection and rescue of critical infrastructure, i.e., the *Law on Protection and Rescue*. It is organized and implemented not only by state and administrative bodies but also by all public institutions, trade companies, including energy operators. According to this Law, protection and rescue is realized through a number of measures and activities, including observation, detection, monitoring and studying potential hazards from natural disasters and other accidents, undertaking preventive measures, reporting and warning, determining and implementing protective measures, supervision of the implementation of protection and rescue, identification and assessment of hazards, and so on. This Law is a systemic ordinance that lists measures and activities that are undertaken to assess and prevent other accidents. The Law defines them as events resulting from certain overlooks and errors in the execution of everyday economic and other activities, as well as negligence in handling hazardous substances and means in the production, storage and transportation of such materials (fires, major road, rail and air accidents traffic, industrial accidents caused by explosions and other technical and technological reasons, spills of oil, and other larger-scale threats). Although critical infrastructure protection is not explicitly stated in the Law, nevertheless it is still recognized in the envisaged measures and activities. As an example, we can cite the legal solution "deliberate overlooks of errors, such as sabotage or diversion" when handling the specified hazardous substances, in which most fall into the primary or final products of a kind of critical infrastructure, that is the energy sector in North Macedonia. One normative item for monitoring and implementing the critical infrastructure protection activities is recognized in the eight principles on which protection and rescue is based, in terms of energy infrastructure. Namely, everyone has the right to protection and rescue from natural disasters and other accidents; the Republic of North Macedonia, municipalities, public enterprises, institutions and services and trade companies are obliged to organize and undertake preventive and operational measures for protection and rescue against natural disasters and other accidents on time; any natural and legal person, in accordance with the Law, is responsible for implementing the foreseen protection and rescue measures, etc. Of particular importance for the protection of energy infrastructure is the principle that obliges the security system institutions, and the public and private sector companies, including the energy operators, to organize and undertake operational measures. Moreover, the strategies for critical infrastructure protection give them a key role in the process of achieving the goals of the respective national strategies. In the *National Strategy for Protection and Rescue*, critical infrastructure protection is perceived from the perspective of protection and rescue planning. The *Protection and Rescue Plan* does not provide a definition of critical infrastructure, although the concept is recognized in explaining the obligations of participants in the protection and rescue system. The *Protection and Rescue Plan* is developed on the basis of the assessment of threat of natural and other disasters, while for the needs of the private sector, including the energy operator, air transport (privately owned), etc., which is based on the available information the Assessment is adopted by a body managing them all. Furthermore, the management also adopts a *Protection and Rescue Plan* based on which undertakes measures and activities to secure the infrastructure against natural disasters and other threats. From the overall analysis of the laws and bylaws, we can conclude that none of the documents in the Republic of North Macedonia provides a complete risk management solution for the functioning of critical infrastructure, as well as a framework for its protection. To date, no law has been drafted on critical infrastructure protection. That is why, the concerned professional and scientific public in North Macedonia has increased the interest in the subject area by organizing scientific debates, workshops and conferences on critical infrastructure. We believe that these initiatives will provide input and initiate political support for the adoption of a law and strategy on critical infrastructure protection.

8.3. Public-Private Partnership

In recent years, public-private partnership has become a key initiative in the Republic of North Macedonia and has gained an increasingly prominent role. Each entity contributes to the specific resources of the system, participates in planning, decision-making and strives to enhance the resilience and protection of critical infrastructure.

The private sector is most often the owner as well as manager of a number of national critical infrastructure: communications, IT sector, etc. (such as Makedonski Telekom, TAV Airports), so it is understandable that the private sector is best acquainted with critical infrastructure requirements – weaknesses and strengths – and must be part of strengthening critical infrastructure resilience and protection. That is why it has the responsibility to protect critical infrastructure and enable day-to-day functioning of all modern entities, national security and international cooperation, as well as the exchange of knowledge, experience and best practices between stakeholders (both private and public), since strengthening resilience and protecting the critical infrastructure system should be even common practice. Therefore, it is quite understandable that the private sector has the responsibility to protect the infrastructure that is important for the functioning of the whole society, and this cannot be done efficiently, effectively and at no cost if there is no cooperation with the public institutions. This cooperation also creates numerous problems in the area of exchange of sensitive information and data, which are crucial for mutual trust, as well as the exchange of knowledge and experience. These problems impose the need to work on further strengthening and greater public-private cooperation. Perhaps by analyzing the technology leader in the country it was revealed that, Makedonski Telekom, which is part of the Deutsche Telekom Group, implements all the procedures, standards and knowledge accepted at all levels by Deutsche Telekom AG Corporation. In addition, the analysis shows that all measures are at a higher level compared to our country, which does not have a single system (law and strategy) for critical infrastructure protection. Threats become more complex and endanger the functioning of infrastructures, which is a major challenge for the state, its bodies and operators. That is why, in an open partnership with the private sector, the public sector can change a lot, striving to better set up and manage the system with higher quality.

The Macedonian normative model of public-private partnership is regulated by the *Law on Concessions and Public-Private Partnership* (2012). This Law prescribes the possibility for contractually regulated cooperation, which, for private security, it would refer to public procurement of a service. The concept of public-private partnership in the private security sector in Macedonia is to some extent applied by the adoption of the *Law on Securing Persons and Property*. This legal solution is limited to investing in the construction and maintenance of the objects and it is not fully adapted to the full needs of critical infrastructure. One way to help overcome this problem is to establish a public-private partnership, specifically cooperation in different areas of critical infrastructure, with clearly defined mutual rights and obligations easily implemented for both parties.

8.4. Cooperation with European Commission

The SAA between North Macedonia and the EU entered into force in 2014. Since 2009, the Commission has recommended to the Council to open accession negotiations with North Macedonia, a candidate country since 2005. Taking into account the progress made in previous years, in April 2018, the Commission reiterated its unconditional recommendation to open accession negotiations and the Council gave the direction to open accession negotiations in 2019. We must emphasize that the Report concludes that North Macedonia continues to follow and deepen reform dynamics, delivering tangible results in the key areas identified in the Council Conclusions from June 2018. However, on the other hand, the Report states that there was no progress in civil protection. North Macedonia continues to participate in the Union's civil protection mechanism, and such a format of cooperation is extremely important regarding policymaking in terms of rational, coordinated and effective disaster mitigation and response at all levels. In December 2018, North Macedonia formally applied to participate in the Hybrid Threat Survey for the EU, in order to identify their key weaknesses and strengthen their resilience to hybrid threats. In January 2019, the survey was launched by the Ministry of Defense. In addition, of particular importance for critical infrastructure protection is the 2018 meeting between the EU and the external partners of the critical infrastructure protection system, where the cooperation was expanded with eight partner countries, including North Macedonia.

The direction of the European Union toward the perception and discussion on critical infrastructure protection is crucial for the development of the policy of the European Union, not only for the Member States but also for the candidate countries for EU membership, such as the example with North Macedonia. The key document adopted in the field of critical infrastructure in the EU is *Directive 2008/114/EC* from 2008, which aims to establish a comprehensive process for critical infrastructure protection at the level of the Member States and at the level of the EU in general. What should be noted is the fact that candidate countries, such as North Macedonia, must implement the very Directive before formally joining the Union. For North Macedonia, the inclusion of the provisions of *Directive 2008/114/EC* in our national legislation is a multifaceted challenge, which means that there is a need for fast adjustment. Furthermore, based on the regulation of the European Union we have the opportunity to develop and implement solutions in favor of Macedonia in building critical infrastructure protection policies. The first step in the multi-phase approach should be directed in identifying and labeling ECI and assessment of the need for improving their protection. Thereby, the focus should primarily be on the energy, transportation and ICT sectors. The second step should be directed toward terminological harmonization of the monitored infrastructure according to the established definition and fulfillment of the cross-sectoral criteria. In this part, it is important for North Macedonia to apply sectoral criteria in order to prepare a primary qualification of critical infrastructure within the respective sector of potential critical infrastructure, to apply the definition of infrastructure, according to which it should be determined whether it meets the requirements and conditions for "critical infrastructure" or "European critical infrastructure". For the realization of all key steps for North Macedonia, the assistance of the European Commission is necessary, starting from the process of creating procedures, the interpretation of certain criteria in overcoming the lack of administrative and professional capacity. Assistance is also needed in the part of adjusting the national framework, that is, in establishing a whole package of programs related to the protection of critical infrastructure, in establishing a normative framework in accordance with *Directive 2008/114/EC* and to discuss with Greece and Bulgaria about possible ECI.

The second important document adopted by the European Parliament and the Council is the *NIS Directive*, which emphasizes the measures for a high common level of security of networks and information systems in the Union. If we start from the main goal of the *NIS Directive*, then we can identify the directions of North Macedonia toward the efforts to ensure a common level of security of networks and information systems that apply to all Member States, whose irregularities caused by security incidents may have strong consequences on society or the economy of the country.

Therefore, for North Macedonia, options remain open for using EU funds for issues related to critical infrastructure. We are still at the beginning regarding the development of critical infrastructure systems, and the start of the accession negotiations is status quo.

8.5. Education and Training

In practice, the need for the same basic knowledge of the conditions and problems associated with critical infrastructure protection is increasingly acknowledged. North Macedonia has modest experience as it is beginning to develop a critical infrastructure protection system and therefore does not have a specific program in any higher education institution. A subject entitled Critical Infrastructure is being studied only at the University of Ss. Cyril and Methodius, the Institute of Security, Defense and Peace, and the Faculty of Philosophy in Skopje. The scientific and expert public has initiated several conferences, seminars, trainings, workshops and exercises related to critical infrastructure protection. To name a few, the Private Security Chamber conducted training on the importance of critical infrastructure (in September 2017) with MoI staff, security agency staff, and managers of security agencies for persons and property. In addition, the Private Security Chamber had been continuously delivering specialized training for physical and technical security of critical infrastructure objects throughout 2018 and 2019. In 2017, the MoI organized an international conference on the protection of critical objects. Case study and simulations give new interaction to the whole process in North Macedonia in line with the new critical infrastructure challenges, and in this regard on 18 July 2019, an exercise was held with the staff of JSC OKTA Refinery (one of the largest manufacturers, distributors, and sellers of petroleum products in Macedonia), organized by the Directorate for Protection and Rescue. In 2019, an exercise was held entitled "National Response 2019 – Managing Major Disasters" organized by the Crisis Management Center, Directorate for Protection and Rescue, Ministries of Defense, Health, Interior, Agriculture, Forestry, and Water Economy, and Red Cross with the main purpose of applying the new Next-Generation Incident Command System (NICS). Publications aimed at presenting and explaining all aspects related to critical infrastructure protection and at the same time initiating activities for strategy development and adopting a Law on Critical Infrastructure Protection in the Republic of North Macedonia are *Critical Infrastructure Guidelines* (Mitrevska & Mikac, 2017) and *Critical Infrastructure: Concept and Security Challenges* (Mitrevska, Mileski, & Mikac, 2019). These two publications have offered a model for developing a critical infrastructure protection system in the Republic of North Macedonia from a scientific, normative, and expert point of view.

Chapter Conclusion

To date, no Law on critical infrastructure protection has been adopted in the Republic of North Macedonia. This area has been regulated within the framework of other laws and bylaws. This chapter analyzes those normative solutions, paying attention to the protection of this area. Without trying to be critical, the analysis shows that a lot of progress has been made, however, opportunities have been missed to pass a special Law on critical infrastructure protection.

The analysis of the situation in the Republic of North Macedonia indicates the necessity for a systematic approach to the existing infrastructure and the need to define infrastructure as critical, due to the possibility to be a potential target. The Republic of North Macedonia should define the most appropriate model for critical infrastructure protection, through regulation and integrated approach, starting from identification, prevention and preparation for tackling threats to critical infrastructure and thus mitigating the consequences on it. The Republic of North Macedonia in parallel with determining the strategic imperatives, it is necessary to provide a good assessment of threats, vulnerability, to establish a list of critical infrastructure and standards for critical infrastructure protection and consequences on it, and above all, to improve resilience, that is, secure infrastructure of possible human, physical and cyber threats.

There are initiatives to adopt a special Law on critical infrastructure protection and it is necessary for them to go through all the stages until its adoption. That is why the need for establishing a normative framework for critical infrastructure protection as a fundamental basis for its identification must be a priority initiative of government institutions. That initiative must be harmonized with other sectoral solutions, strategies, laws, programs and plans. Therefore, through an authorized ministry, the government needs to play a key role and be the coordinator of the entire system and propose a Law on critical infrastructure protection. A list of critical infrastructure needs to be formally established. A formal framework for the protection of national critical infrastructure needs to be adopted, as well as a strategy and a law, which would normatively commence the establishment of a critical infrastructure protection system. The Law should emphasize the place and role of public-private partnership, education, and training. However, above all, among the most important activities should be defining facilities as critical infrastructure, defining measures for their protection and security, and defining duties and responsibilities of the institutions. This process should be in line with the guidelines for normative regulation of issues in the field of identification, designation and protection of ECI, with special emphasis on the obligations arising from the *Council Directive 2008/114/EC.*

Notes

1. This Decision was adopted by the Government of the Republic of Macedonia in 2013, and the need for its adoption derives from the Law on Private Security from 2012 and the Law on Amending and Supplementing the Law on Private Security, adopted in 2013.
2. Decision for designating the legal entities obliged to private security, "Official Gazette of the Republic of Macedonia", no. 106/2013, article 2.

3. Aviation Law, Official Gazette of the Republic of Macedonia, Skopje, no. 63/2015.
4. Alcheski, G J. (2016), Implementacija na sovremenite bezbednosni sistemi i procedure vo razvojot na obezbeduvanjeto na objektite od vitalen interes za Republika Makedonija (so osvrt na aerodromskata bezbednost), Filozofski fakultet, Skopje, pp. 213–214.
5. This Decision was adopted by the Government of the Republic of Macedonia in 2013, and the need for its adoption derives from the Law on Private Security from 2012 and the Law on Amending and Supplementing the Law on Private Security, adopted in 2013.
6. Decision for designating the legal entities obliged to private security, "Official Gazette of the Republic of Macedonia", no. 106/2013, article 2.

Chapter 9

Albania

Stavri Dhima, Sara Prenga and Ardi Dhima

Contents

As a country that is in an important stage of development and modernization of the infrastructure in the energy and transport sector, which in essence constitute the critical infrastructure of any country, Albania has accompanied this development and modernization of its infrastructure with the development of networks and secure information systems. In support of this development, Albania has adopted and approved the framework of relevant legislation, as well as approved the relevant policy documents and strategies for the protection of critical infrastructure, as well as cyber security in the country. Achieving these results has been accomplished through reliance on EU legislation and policies in these two areas, as well as by aligning the Albanian legal and institutional framework with that of the EU.

In achieving the objectives regarding the protection of critical infrastructure, as well as cyber security in the country, the responsible institutions and structures in these sectors have taken into account that the security of activities in these areas (of infrastructure and information) is not only a basic requirement for a sustainable economic and social development, but they also depend on the context of regional

DOI: 10.4324/9781003273769-11

and international developments, as well as the geostrategic position of Albania in this context. Such an assessment is also defined in the *National Security Strategy 2014–2020*, approved by *Law No. 103/2014 "On the approval of the National Security Strategy of the Republic of Albania"* (Albanian Assembly, 2014a), where in point 3, of chapter II, of this Strategy, it is stated that "The geostrategic position of Albania is one of the most important assets for the economic development of the country. . . . The selection of the Trans-Adriatic Pipeline (TAP) is a clear indication of the advantages of this position."

9.1. Introduction

Albania, being a developing country, which aims at full integration in Euro-Atlantic structures, both in terms of economic development and in terms of security, defense and the rule of law institutions, has aimed at achieving and consolidating a sustainable economic development, which has, as preconditions, a developed and modern infrastructure (in energy and transport) as well as a modernization to contemporary standards of information networks (cyber networks). But in the meantime, the requirements for sustainable development are not only related to the development and modernization of transport and energy infrastructure as well as that of information but it is also required to guarantee a safe operation of these infrastructures in terms of business activity, as well as in terms of technical and technological security of the respective assets and installations, which is achieved through the establishment and operation of relevant supervision and inspection of institutions whose activity is based on a complete legal framework according to European standards. It is precisely for this reason that energy infrastructure and transport infrastructure fall into a specific category such as critical infrastructure (CI), for which legal and institutional protection is required, which is also required for cyber infrastructure.

However, the implementation of sectoral strategies in every field, especially in energy and transport, is closely linked and conditioned by the development of information networks (cyber networks); therefore, it is naturally accepted that the strategies for developments in the energy and transport sectors, should be harmonized and in synergy with the cyber security strategy. In this concept, in point 9, of chapter II, of the *National Security Strategy 2014–2020* (Albanian Assembly, 2014a), it is defined as a very important objective and

> Approval and implementation of the National Strategy for Cyber Security, for setting and respecting the highest standards in direction of preservation and protection of information in all forms of its existence, while focusing special efforts on the protection against cyber-attacks.

Taking into consideration the Euro-Atlantic integration processes where Albania is currently involved (Albania has been a member of NATO since 2009, and is in the phase of opening EU integration negotiations), it is natural that developments in such strategic sectors as energy and transport with their respective infrastructure, but as well the IT sector, be gradually integrated at the European level.

This integration process on a regional and European level, which has been a priority of Albania for the last 30 years, in terms of energy and transport sectors was concretized with the participation of Albania in the Treaty for the Establishment of the Energy Community (EnC), signed on 25 October 2005 (ratified by the Assembly with *Law No. 9501, "On the ratification of the Treaty establishing the Energy Community"*) (Albanian Assembly, 2006a), as well as by participating in the Treaty establishing the Transport Community (TeC), signed on June 2, 2017 (ratified by the Assembly with *Law No. 8/2018 "On the ratification of the Treaty establishing the Transport Community"*) (Albanian Assembly, 2018).

Referring to the integration processes, as well as the obligations arising from the fact that the country is already a contracting party to both of these treaties, which also define the *Acquis Communautaire* which are binding on each member country to implement these treaties, Albania has built a legal and institutional framework at a European level for the development of the energy and transport sector, as well as for IT (Albanian energy legislation is based on the EU's Third Energy Package).

The same processes have occurred in terms of approximation of domestic legislation with that of the EU in the field of protection of CI, as well as security of information networks (INs). Albania has partially approximated *Council Directive 2008/114/EC* "on the identification and designation of European critical infrastructures and the assessment of the need to improve their protection" (Council of the European Union, 2008), not through a special law but as part of *Law No. 45/2019 "On civil protection"* (Albanian Assembly, 2019).

Meanwhile, with *Law No. 2/2017* "*On cyber security*" (Albanian Assembly, 2017), is partially approximated the *Directive (EU) 2016/1148* of the European Parliament and of the Council, dated 6 July 2016, "concerning measures for a high common level of security of network and information systems across the Union" (European Parliament and of the Council of the European Union, 2016).

Also, the Government of Albania has prepared and approved a series of sectoral and cross-sectoral acts and strategic documents in support of the fight against terrorism, seeing this fight in a broad perspective, including intertwining with the fight against organized crime and various forms of financing of terrorist acts. Thus, *Law No. 157/2013 "On measures against the financing of terrorism"* (Albanian Assembly, 2013a), as well as *Decision of the Council of Ministers (DCM) No. 663, "On the approval of the intersectoral strategy of the fight against organized crime, illegal trafficking and terrorism, 2013–2020 and the action plan for the years 2013–2016"* (Council of Ministers, 2013).

However, the legal and institutional framework in force regarding CI and its protection, as well as infrastructure networks (INs) security, needs to be improved, updated and harmonized with issues of risk and security management and sustainability at a regional and European level, as these are directly related to the performance of other sectors of the economy, and to the overall socio-economic progress of the country.

9.2. Implementation

According to the legal framework in force, the legal and institutional administration and management of the CIP system in Albania is realized through the provisions of *Law on civil protection*, which is a very important law for setting up of a national system for the protection of CI.

Law on civil protection represents a general legal act for the assessment, prevention, protection and management of natural disasters (point 9 of article 3 of the Law, "Definitions"), but also of other disasters (point 10 of article 3 of the Law), in order to have a legal and institutional harmonization of the treatment and, above all, of the protection of the CI.

In this Law, CI and its protection are treated in a special chapter, chapter 5. Article 43 of chapter V, states what constitutes the CI, and specifically point 1 of this article expresses and identifies the CI that consists of sectors and cross-sectors, within 10 sectors recognized by Law. Point 2 of article 43 defines the responsibilities and duties of public institutions, central and local structures, or private entities that have in administration or critical CI.

Given that infrastructures in general, such as energy and transport infrastructure, are not limited and contoured only within a country, but they have regional and even intercontinental dimensions and scope, it is very important to assess and manage CI with cross-border effect (article 44), which also defines the tasks for the ministries, which are given according to point 1 of this article, while according to point 2 of article 44, the Albanian state informs other states that may be affected by the interruption or destruction of its CI by the cross-border effect.

Under the provisions of *Law on civil protection*, currently, the Ministry of Defense, as the ministry responsible for civil protection, is the ministry responsible for the implementation of this Law, whose responsibilities and duties are defined in article 21 of the law, meanwhile, The National Agency for Civil Protection (NCPA) also functions as part of the structure of this ministry. The Ministry of Defense is also the ministry that determines the strategic directions and targets of NCPA. In more detail, the responsibilities of NCPA are given in article 23 of *Law No. 45/2019*, meanwhile, as defined in point 4 of article 23 of the Law, the organization and functioning of NCPA are defined through *DCM No. 747 "On the organization and functioning of the National Agency of Civil Protection"* (Council of Ministers, 2019a).

Since many of the dangers and accidents that threaten CI are related to fires, article 36 of the *Law on civil protection,* defines the role and responsibilities of the Fire Protection and Rescue Service (FPRS). However, the organization and management of the FPRS that faces extremely dangerous situations in cases of natural disasters, as well as in cases of other disasters (including impact on CI), is defined in a special law, in *Law No. 152/2015 "On the Fire Protection and Rescue Service"* (Albanian Assembly, 2015a), as well as in *DCM No. 520 "On the approval of the regulation for the Fire Protection and Rescue Service"* (Council of Ministers, 2019b).

In Albania, the process of identification, priority, protection, sustainability and full legal regulation in the field of CI still requires to be institutionalized despite the fact that the main part of the energy and transport infrastructure, which belongs to objects of economic and state importance, is included in the framework of *DCM No. 288 "On determining and taking measures against fire and rescue in buildings of economic and state importance"* (Council of Ministers, 2002) (where in its annexes 1 and 3 are included mainly objects of CI in the field of energy and transport), a decision which should be reviewed, updated and supplemented.

As discussed earlier, in the framework of cooperation in the region of South-Eastern Europe (SEE), which includes Albania, the EU has conceived the form and level of setting up specific policies and relations with the countries of the Enlargement Agenda, through the adoption of the Treaty establishing the EnC, as well as the Treaty establishing the TeC. Participation in these two treaties has made possible the institutionalization in the field of CI of EU cooperation with neighboring countries in the SEE region, which will join the EU. In the energy sector, the obligations arising from the Energy Treaty have materialized with the implementation of the *Acquis Communautaire* through the adoption of the *Law No. 43/2015 "On the electricity sector"* (Albanian Assembly, 2015b), as well as the *Law No. 102/2015 "On the natural gas sector"* (Albanian Assembly, 2015c).

Based on the document of the National Energy Strategy for the period until 2030 approved by *DCM No. 480 "On the approval of the national energy strategy for the period 2018–2030"* (Council of Ministers, 2018a), which is designed to meet the development needs of the country and to provide its citizens with a sustainable, reliable and affordable energy supply based on regional integration and diversification of the energy sector, of special importance for this sector is the energy integration in the region and in the EU, which will be a crucial step for Albania to meet its growing demand for energy, but also to be a factor in the energy market in the region and beyond.

Albania is considered a critical node in the creation of two possible corridors of energy market integration in the region of Southeast Europe, intensified by the development of strategic projects of LNG terminals; these strategic projects include the Adriatic Integration Corridor and the Southern Integration Corridor through Ukraine.

Given that the protection of CI requires very specific and well-organized decisions and actions, it is very important to know and identify the extent of this infrastructure, its interconnection, both for energy and for transport infrastructure, which by their nature have a cross-border extent and effect, as treated and reflected according to the *General National Plan of the Territory* approved by *DCM No. 881 "On the approval of the General National Plan of the Territory"* (Council of Ministers, 2016a).

Interconnection of the regional natural gas network with the foreseen natural gas transmission network in Albania is given according to the Gas Master Plan for Albania, approved by *DCM No. 87 "On the approval of the development plan of the natural gas sector in Albania and the identification of priority projects"* (Council of Ministers, 2018b).

Assessing the treatment given earlier for energy infrastructure, the same treatment is done in terms of transport infrastructure, which is also part of the CI and by nature has a cross-border effect. The current situation and the expected development of the critical transport infrastructure in the country and its interconnection at the regional level and beyond is also given in the *General National Plan of the Territory*, which includes road transport lines and multimodal junctions, together with the continuation of this infrastructure in neighboring countries in the region, as well as railway transport lines, multimodal junctions and the connection of railway infrastructure with neighboring countries in the region.

Current developments in the transport sector and relevant infrastructure have a general target, which is defined in the Transport Sector Strategy 2016–2020, approved by *DCM No. 811 "On the approval of the Transport Sector Strategy and action plan 2016–2020"* (Council of Ministers, 2016b). In the document of the *Transport Sector Strategy 2016–2020*, in point 1.1.2., "Albanian transport sector in the regional and EU context", a reference of this Strategy is made in the "European Transport Policy".

Another important law for the identification of CI is *Law No. 107/2014 "On territorial planning and development"*, as amended (Albanian Assembly, 2014b), based on the General National Plan of the Territory, followed by regional and local plans. Of particular importance are the spatial plans of areas of national importance, which generally coincide with CI.

Despite the fact that the *Law on civil protection* has partially approximated the *Council Directive 2008/114/EC*, the "object" of *Directive 2008/114/EC* is not fully covered by *Law on civil protection*, which must be corrected with the necessary legal improvements, including the possibility of a new separate law.

An essential aspect of CIP is the "assessment of the need to improve the protection of (critical) infrastructure", which includes the prevention and risk assessment of CI, through control and institution inspections in relevant areas, such as the State Technical and Industrial Inspectorate (STII) for the oil, gas, electricity and under pressure vessels sector, which is an institution that has been established and operates based on *DCM No. 410 "On the establishment, organization and functioning of the State Technical and Industrial Inspectorate"* (Council of Ministers, 2015a), as well as the National Inspectorate of Territorial Protection (NITP) in terms of transport infrastructure in general, an institution that is established and operates on *DCM No. 893 "On the organization and functioning of the National Inspectorate of Territorial Protection"* (Council of Ministers, 2015b).

It is very important that the role and functioning of STII and NITP, in terms of prevention and risk assessment for CI, be harmonized and integrated within the structures and institutions that are established and operate pursuant to *Law on civil protection*. In this context, it will be necessary that *DCM No. 965 "On inter-institutional cooperation of management structures, in cases of civil emergencies and crises"* (Council of Ministers, 2015c), be reviewed to be adapted and harmonized more effectively with the *Law on civil protection*, and more specifically in relation to *Council Directive 2008/114/EC*, on the CIP.

Meanwhile, in increasing productivity in INs classified as critical, which have an impact on the economic and social development of the country, as well as in many other areas, an important role is played by ICT, whose massive use is currently a major factor in these developments. Online security of CI, especially in the energy sector, is becoming more important for security and assurance of production, distribution, transmission and storage of energy as well as the stability of a single European energy market.

Permanently, most of the CI like that of energy systems, as well as the infrastructure of transport systems depend on the efficient functioning of the communication networks. These support systems and networks, commonly referred to as CII, constitute the main pillars for the functioning of the economy and society. An attack or disruption to these assets and services has a direct impact on the country's economy as well as to a large extent of the population.

Albania ranks among the countries where the development of telecommunications, internet access and computerization of society is progressing very fast, but, at the same time, it exposes it to the dangers of cyber nature with state and non-state actors. Also, with NATO membership and the progress made toward EU membership, Albania is becoming a more active part of cyber security initiatives and programs.

In accordance with the commitments of the SAA, during the last two decades, Albania has drafted and adopted a number of important laws for electronic communications in accordance with the EU *acquis* such as *Law No. 9918 "On Electronic Communications in the Republic of Albania"*, (Albanian Assembly, 2008), amended by *Law No. 102/2012*, as well as specifically the approval of *Law on cyber security*, which has partially approximated *Directive (EU) 2016/1148* of the European Parliament and of the Council.

According to article 1 of *Law "On cyber security"*, its goal is to achieve a high level of cyber security, defining security measures, rights, obligations, as well as mutual cooperation between entities operating in the field of cyber security. Whereas, in points 5 and 6 of article 3 of this Law, there are given the definitions for "Important information infrastructure" and "Critical information infrastructure".

National Security Strategy 2014–2020 (Albanian Assembly, 2014a), in Appendix C – "Risk Classification", defines cyber attacks as a first-level risk. The Republic of Albania has a cross-sectoral approach to internet security, involving numerous state agencies, but the National Authority for Electronic Certification and Cyber Security (NAECCS) is the national authority responsible for internet security.

Pursuant to point 2 of article 6, of *Law "On cyber security"*, the Council of Ministers, upon the proposal of the responsible minister, approves the list of CIIs and important information infrastructures, which is updated at least once every two years. Pursuant to these legal requirements have been adopted *DCM No. 553 "On the approval of the List of Critical Information Infrastructures and the List of Important Information Infrastructures"* (Council of Ministers, 2020).

Meanwhile, Energy Regulatory Entity (ERE) has approved *Decision No. 126 "On the approval of the "Regulation on cyber security of critical infrastructures in the electricity sector"* (Energy Regulatory Entity, 2020). Referring to this Decision of the ERE, with the proposal of the operators in the electricity sector, it is requested that part of the CI be:

*"**OSHEE** proposes . . .: The billing system of OSHEE; Meter reading system (Mobiread); Financial Systems (SAP); Substation and cab monitoring system (SCADA)*

***OST** proposes . . .: Supervisory Control and Data Acquisition (SCADA)*

***KESH** proposes . . .: ABB Symphony SCADA system; Data transmission (communication) network of KESH (Tirana – Vau Dejës HPP, Tirana – Fierza HPP, Tirana – Komani HPP)"*

These proposals have been evaluated and have become part of the "List of Critical Information Infrastructures", according to Annex 1 of "On the Approval of the List of Critical Information Infrastructures and the List of Important Information Infrastructures".

As for transport (air, sea, road), part of the "List of Critical Information Infrastructures", according to Annex 1 are included:

***ALBCONTROL:** Air Navigation Services of Albania*

***Durrës Port Authority:** The control system of the entry-exit regime in the Port of Durrës; Terminal Operating System (TOS); VMS system of cameras in the territory of Durrës Port.*

***General Directorate of Road Transport Service:** eDM system; eDPSHTR system.*

Except NAECCS, which was established and operates based on *DCM No. 141 "On the organization and functioning of the National Authority for Electronic Certification and Cyber Security"* (Council of Ministers, 2017), there are two other institutions working in the field of Internet security: the Ministry of Defense and the State Police and prosecutor's office Cybercrime Investigation Unit. The Cybercrime Investigation Unit is designated as a single PoC under the Budapest Convention.

In accordance with the requirements of the legislation in force, the Government of Albania has approved *DCM No. 973 "On the approval of the cyber security policy document 2015–2017"* (Council of Ministers, 2015d). This document is also in line with the Digital Agenda for Europe 2020, as well as the Cyber Security Strategy of the European Union: open, secure and protected cyberspace. In accordance with what is defined in this document, in the following period the Albanian Parliament has approved *Law on cyber security*.

In order to promote a high level of security in information systems and critical INs, the Energy Community (EnC) has adopted the Procedural Act in order to facilitate strategic cooperation and information exchange. In accordance with this objective, the Secretariat of the EnC has conducted the "Study on Cyber Security in the Energy Sector of the Energy Community" (Blueprint Energy Solutions GmbH, 2019), (which analyzes the situation of cyber security in all Contracting Parties of the Energy Community), with the aim of contributing to the building of specific internet security capabilities at the national and EnC level. In order to achieve this goal, the Cyber Security Coordination Group operates within the EnC, in the activities of which all contracting parties of the EnC participate, including Albania.

9.3. Public-Private Partnership

During the years of transition in Albania, various governments have made continuous efforts to improve and develop the infrastructure, as well as the provision of public services that promote economic growth. Due to a large inherited infrastructure gap in the public sector, the potential of Public-Private Partnerships (PPPs) as a source of alternative funding was easily accepted by the Albanian governments, which made numerous efforts to create a suitable environment for PPPs in the field of transport and energy infrastructure. In this case, the treatment of CIP through PPPs is very important, which in essence can be in the form of technical and physical protection of critical infrastructure, but also through the protection of INs connected to this infrastructure.

Albania also has a proactive approach in the field of INs (internet) security, including the energy sector, creating a comprehensive legislative and institutional framework to address internet security, as well as establishing a mechanism for identifying relevant issues. Cooperation between stakeholders in the energy sector does not differ from the public or private sector, as *Law "On cyber security"*, or *DCM*

No. 553 "On the approval of the List of Critical Information Infrastructures and the List of Important Information Infrastructures" (Council of Ministers, 2020), does not distinguish between the public and private sectors, nor does it detail which organizations/entities may be involved in this field.

The first PPP efforts made possible by legal reform were made in the 2000s. After that, further legal reforms were introduced, such as the main law governing PPPs in Albania, *Law No. 125/2013 "On concessions and public-private partnership"* (Albanian Assembly, 2013b), which was essentially amended with the approval in 2015 of *Law No. 77/2015* (Albanian Assembly, 2015d).

The current PPP law, with amendments, which constitute the legal and institutional basis of the activity and procedures for concessions and PPPs, clearly defines the legal framework for PPPs and the institutional responsibilities, as this law approximates the *Directive 2004/18/EC* of the European Parliament and of the Council of March 31, 2004, on the coordination of procurement procedures for the awarding of public contracts for works, supplies and services (European Parliament and of the Council of the European Union, 2004).

The legislation has defined four general types of PPP contracts: a) public works concession, b) public service concession, c) public works contract and d) public service contract. The concession/PPP contract is concluded for a certain period, which initially cannot be tied for a period longer than 35 years.

In the period after the adoption of *Law "On concessions and public-private partnership"*, and its subsequent changes, a lot of infrastructure projects belonging to the CI (in both energy and transport) are implemented through concession/PPP-based contracts such as electricity production, transmission and distribution systems, production, refining, treatment, storage and distribution of gas through pipelines, oil production and its byproducts, storage and distribution through pipelines, telecommunications (networks, systems), and transport systems (fuel supply, railway network, road network, airports, ports, inland transport). The Albanian government has awarded PPP contracts for transport infrastructure to several companies for the maintenance and operation of the Milot–Morine highway (Albania to Kosovo highway) as well as for the contraction, maintenance and operation of Milot–Balldren and Orikum Llogara road projects.

For high management and effectiveness of PPPs, it will be necessary to increase the supervisory and controlling role of public financial institutions, mainly the supervisory and controlling role of the Ministry of Finance and Economy, as well as other control inspection institutions.

9.4. Cooperation with European Commission

Albania is a candidate country for EU membership by decision of the European Council in June 2014, while the SAA entered into force in April 2009 (*Law No. 9590 "On the ratification of the Stabilization and Association Agreement between the Republic of Albania and the European Communities and their member states")* (Albanian Assembly, 2006b) which is reflected in the fact that one of the strategic objectives of the country is regional cooperation at a European level, an objective set out in all National Policy Documents for the security of CI (energy and transport), as well as INs.

In accordance with the implementation of commitments within the EU membership processes, in 2018 the *National Plan for European Integration 2018–2020* (NPEI) was approved by *DCM No. 246 "On the approval of the National Plan for European Integration 2018–2020"* (Council of Ministers, 2018c).

Energy and transport issues are addressed in chapter 14 and chapter 15 of the NPEI 2018–2020, while infrastructure issues in the context of Trans-European infrastructure are addressed in chapter 21 "Trans-European Networks". Given how energy and transport infrastructure constitute CI sectors, these chapters also cover issues related to the protection of these infrastructures.

According to article 107 of NPEI 2018–2020, the cooperation in the energy sector will be based on the Regional Treaty establishing the EnC signed with the aim of gradual integration of Albania in European energy markets. Chapter 10 of NPEI 2018–2020, "Information society and media", addresses the issues of INs and their protection. In this chapter, it has been considered that article 104 of the SAA deals with the cooperation that focuses on electronic communications, networks and services, including the protection of these networks.

As a NATO member country, since 2009, and as a country in the process toward EU membership, Albania has institutionalized the forms of cooperation with the countries of the region within the EU integration processes. Referring to two important and strategic sectors, such as the energy sector and the transport sector, Albania is a contracting party to the Treaty establishing the EnC, as well as a contracting party to the Treaty establishing the TeC.

Considering the cooperation with the EU as a supporting and contributing factor in the development of the energy sector, as well as the transport sector, Albania has based its legal framework in these sectors on the gradual transposition of the main EU Directives. Albania participates at CIP meetings and workshops that EC organizes with external partners (as part of EPCIP) for deepening possibilities of cooperation.

The same has been done in terms of CI and its protection (energy and transport), through the partial adaptation of *Council Directive 2008/114/EC* (*Law on civil protection*), as well as through the partial approximation of *Directive (EU) 2016/1148 of the European Parliament and of the Council* (*Law "On cyber security"*).

Through IPA funds, the EU has provided funding to support the preparation of projects in the field of infrastructure, as well as funding for the realization of the construction of this infrastructure. (In the last decade, the concrete contribution was the financing within the Western Balkans Investment Framework initiative –WBIF).

9.5. Education and Training

To ensure adequate protection of CI as well as information/cyber networks, the key factor is the strategic organization and management that will need to be done by the state through relevant human resources. To enable the achievement of the most advanced standards in the protection of CI systems, as well as cyber networks, and for this to be realized by all stakeholders of the public and private sectors in order for them to carry out the necessary activities, it is necessary that the responsible structures and relevant experts have basic knowledge regarding these issues.

The basic legislation related to the definition and protection of CI systems, as well as that of the security and protection of cyber networks, sets clear tasks related to education and training in these areas. Thus, in the legislation that also deals with CI, *Law on civil protection,* in its chapter 10 in article 62, "Education", is stated that "pre-university education provides knowledge on disaster protection and prevention, in accordance with their profile." Meanwhile, in point 2 of the same article, concrete tasks are defined for the ministry responsible for education, which in cooperation with NCPA, should ensure the inclusion in the curriculum of knowledge for civil protection, and of course including knowledge for the protection of CI as well as cyber networks. More specifically, the manner of organization and functioning of higher education institutions, scientific research institutions and various public and private entities are defined in article 52.

However, regardless of what the applicable legal framework stipulates regarding education and training in the protection of CI and INs in higher education institutions, there is still no specific comprehensive program through which all those who are engaged in activities related to CI can gain the necessary education and qualification for the basic knowledge necessary to identify, define, protect and strengthen the sustainability of CI and INs.

In the main institutions of higher education in Albania, which include education programs and curriculums for technical and vocational training in terms of design, construction, operation and protection of CI and INs (such as the Polytechnic University of Tirana, EPOKA University, as well as some faculties of the University of Tirana), it is necessary to include concrete tasks in creating opportunities for education, qualification and training of professional experts capable of preventing the risk of disruption of CI activities, as well as being able to respond when required by NCPA.

In recent years, institutions and experts engaged in the sectors of energy infrastructure, and INs have attended and participated in many national and international activities in these fields, as well as training seminars that have been organized on these issues. Thus, in November 2014, representatives of the Ministry of Infrastructure and Energy, as well as the INs sector, participated in the "NATO Advanced Research Seminar" on "Protecting critical energy infrastructure against new challenges" organized by the Atlantic Council in Tbilisi, Georgia. In this seminar, Albanian experts presented the papers "Protection of critical energy infrastructure against new security threats" as well as "Trans-Adriatic Pipeline – Focus and Impact in Albania" (NATO, 2015).

Another activity was the workshop organized by the OSCE in Tirana on the April 26 and 27, 2017, which addressed issues related to the impact and interdependence of CEI issues with cyber security issues.

In terms of education and training in the field of CI and its protection, as well as information infrastructure protection, for all stakeholders in the issues of these infrastructures, the initiatives for publications related to CIP and the protection of cyber networks will be of great interest, as well as the translation of publications with general values in the field of security and protection of CI and also of cyber security, especially by countries in the region that already have a more consolidated experience in this field.

Chapter Conclusion

In the field of CI and its protection as well as the protection and security of INs (cyber infrastructure), Albania has made significant progress in establishing regulatory acts to guide them. Specifically, through the approval of *Law on civil protection* the *Council Directive 2008/114/EC* has been partially adopted. The same result has been obtained through the adoption of the *Law "On cyber security"* which has partially transposed the *Directive (EU) 2016/1148* (so called NIS Directive) into the national framework. Meanwhile, Albania has supported EU policies and initiatives for a continued rapprochement with the EU, strengthening integration processes through greater regional cooperation and integration, especially in the energy and transport sectors and related infrastructure. In this context, in 2005, the Treaty establishing the EnC was signed (ratified by the Assembly in 2006), while in 2017, Albania signed the Treaty establishing the TrC (ratified by the Assembly in 2018).

In Albania, the process of identification, prioritization, protection, sustainability and full legal regulation in the field of CI still requires further improvement and institutionalization, including institutions and other structures of control and inspection of energy infrastructure and of transport (such as STII and NITP) coordinated with other institutions. Also, regarding the energy and transport infrastructure, which are considered facilities of economic and state importance, within *the DCM No. 288 "On determining and taking measures against fire and rescue in buildings of economic and state importance"* (Council of Ministers, 2002), changes and additions are required in its annexes 1 and 3 where all major CI facilities in operation must be included.

Also, referring to *Council Directive 2008/114/EC*, but also specifically to the provisions of article 10 of *Law on civil protection*, Albania must approve the *National Strategy for Disaster Risk Reduction*, which must be approved by the Council of Ministers on the proposal of the Minister responsible for civil emergencies (and reviewed every five years), and must approve the National Plan of Critical Infrastructure and its protection for a period of up to five years, taking into consideration the very positive experience in this regard that other countries in the region have, such as Croatia and Kosovo. This document should also include the priorities of capacity-building activities and training to implement key aspects of the CIP legal and institutional framework.

In this context, it will be necessary for the European Commission to support national or regional projects for the protection of CI in order to support Albania and other countries in the region in the consolidation and further development of the necessary normative framework, creating platforms for cooperation between countries in the region for the protection of CI with regional impact and building the necessary human and technological capacities for the implementation of the CIP Framework.

Considering the partial approximation of *Council Directive 2008/114/EC*, (through *Law on civil protection*) as a very important achievement, it would still be considered more effective that the transposition of this Directive, be done through a special law, defining concretely as the responsible institution the ministry responsible for infrastructure and energy, as well as the establishment of relevant structures to monitor the implementation of legislation on CI.

Meanwhile, regarding the security of cyber infrastructure (INs), the legal framework can be considered almost complete with the adoption of *Law "On cyber security"*, and bylaws issued in implementation of this Law, while NAECCS is established and functional, which monitors the implementation of security measures by operators of CI and important INs, through control procedures.

Based on the standards and best practices to ensure cyber security, Albania will have to take all measures to consolidate and implement policies, standards, guidelines and procedures to ensure protection against cyber threats, in compliance with fundamental rights and freedoms as well as other democratic principles.

Strategies and programs for CI and its protection, as well as for cyber security, should be adapted to the needs and readiness of the country to implement them, including the consolidation of national capacities in these areas, without avoiding the reflection of the future needs of the country.

Chapter 10

Kosova

Dardan Koçani and Alessandro Lazari

Contents

This chapter provides an insight into the progress that the Republic of Kosova has made in developing its normative and strategic framework on critical infrastructure protection and cyber security, both of which are largely in line with the EU acquis and policies. It also provides an overview of the *National Critical Infrastructure Protection Plan for the Republic of Kosova 2020–2023* which could be considered as a good example for other countries planning to develop the normative framework on CIP in line with the European Union (EU) legislation and policies.

10.1. Introduction

The Republic of Kosova has transposed the *Council Directive 2008/114/EC of 8 December 2008 on the identification and designation of European critical infrastructures and the assessment of the need to improve their protection on critical infrastructure*

DOI: 10.4324/9781003273769-12

(hereinafter *Council Directive 2008/114/EC*) by approving the *Law on Critical Infrastructure* in 2018 which entered into force in April 2019. On the other hand, in 2019 the MoI initiated the drafting of the *Law on Cyber Security* in line with two Directives: 1) *Directive (EU) 2013/40 of the European Parliament and of the Council of 12 August 2013 on attacks against information systems replacing Council Framework Decision 2005/222/JHA* (hereinafter Council Directive 2013/40); and 2) *Directive (EU) 2016/1148 of the European Parliament and of the Council of 6 July 2016 on security measures for network and information systems* (hereinafter *Directive (EU) 2016/1148*). In addition, the government in 2019 approved the *Concept Document*[1] *on Information Systems and Networks Security Measures* in order to transpose *NIS Directive.*

In order to develop the necessary legislative and institutional framework, the MoI initiated the drafting of the *Concept Document on Identification and Protection of Critical Infrastructure.* The *Concept Document*, which was approved by the government in 2015, recommended developing a law on national critical infrastructure in line with *Council Directive 2008/114/EC.* This was the first comprehensive policy that laid down the foundations of the Law on Critical Infrastructure by embracing the all-hazard approach. The *Concept Document* was based entirely on *Council Directive 2008/114/EC* and the relevant policy papers produced by the European Commission in the context of the European Programme (EPCIP). The goal of the *Concept Document* was to develop a law that would regulate the protection of national critical infrastructure and establish a relevant institutional framework. Thus, the procedures foreseen in the *Council Directive 2008/114/EC* for identifying and designating ECI were transformed into procedures for the identification and designation of national critical infrastructures. The drafting of the *Law on Critical Infrastructure* began in 2017 under the lead of the MoI which established an inter-institutional working group. The *Croatian Critical Infrastructure Law* together with *Council Directive 2008/114/EC* was used as the main reference documents.

However, there was an issue in transposing *Council Directive 2008/114/EC* into national legislation: the Republic of Kosova is not a member of the EU and the *Directive* provided obligations to member states in identifying and designating ECI. For the Government of the Republic of Kosova, the protection of national critical infrastructure was of utmost importance, however, the regional context was important as well. Critical infrastructures, namely in the energy and transport sector, are located in countries of the Western Balkans whose disruption or destruction would have significant impacts on the entire region. In addition, considering that Western Balkan countries are unlikely to become members of the EU in the near future, the region could not afford a lack of coordination to protect critical infrastructure with regional impact. For this reason, the Kosova *Law* defined ECI as *a classification of critical infrastructure whose disruption or destruction would have a significant impact on at least two (2) European countries.* The novelty was to use the word *European countries* in order to include the Western Balkans, EU Member States and other European countries; therefore, keeping the national law in line with the need to look into transboundary externalities, which was introduced by the *Council Directive 114/08/EC.*

Overall, both EPCIP and Cybersecurity policy packages are reflected in Kosova's national framework in three main policy areas: counter-terrorism, cyber-security and emergency response.

In the field of counter-terrorism, the Government of the Republic of Kosova has approved three national strategies[2] in line with the *EU Counter Terrorism Strategy*, and all of them have prioritized the protection of critical infrastructure from terrorist attacks, however, they lacked a fundamental aspect – critical infrastructure was not identified and designated. Despite this gap, Kosova Police and other institutions from the security sector have continuously developed and enforced operational plans to protect infrastructures from potential terrorist attacks.

In the cyber domain, in 2015 the MoI launched the development of the first *National Cyber Security Strategy* in accordance with European Union Agency for Cybersecurity's (ENISA) methodology. The *National Cyber Security Strategy and Action Plan 2016–2019* was approved in January 2019 and has identified five strategic objectives: 1) CII protection, 2) institutional development and capacity-building, 3) building public and private partnership, 4) incident response, and 5) international cooperation. The *Strategy* envisaged the identification of CII assets and services according to ENISA methodologies; however, this objective was not accomplished. In addition, the *Strategy* identified two legislative acts that needed to be approved: the *Law on Critical Infrastructure* and the amendment of the *Law on Preventing and Fighting Cybercrime.* The *Law on Critical Infrastructure* and the *Concept Document on preventing and fighting cybercrime* were both approved in 2018. This *Concept Document* recommended developing a new *Law on Cyber Security* in accordance with *Council Directive 2013/40* and *Directive (EU) 2016/1148.* The MoI in 2019 initiated the drafting of the *Law on Cyber Security* and the new *National Cyber Security Strategy and Action Plan 2020–2025*, which are still not approved by the time of publishing this book.

In addition, the Government of the Republic of Kosova in 2019 approved the *Concept Document on Information Systems and Networks Security Measures* which was sponsored by the Ministry of Economic Development. This document recommended to develop a *Law on Security Measures for Networks and Information Systems* and transpose the *NIS Directive.* Moreover, it provided to the Ministry of Economic Development the authority to monitor and coordinate the activities on the security measures of NIS used as critical infrastructure for the private and public sectors. However, it is still unclear if the new *National Cyber Security Strategy* and *National Security Strategy*[3] will clarify the institutional, policy and legislative architecture of both cyber security and CIP.

In the field of EM, the government, in 2010, approved the *National Response Plan* and *Integrated Emergency Management System* which provide specific measures and plans for management and response to incidents. They implement the *all-hazard approach* and define the roles and responsibilities of the government, non-government organizations and the private sector in managing national incidents. However, both documents need to be reviewed as the security sector architecture has changed in December 2018 with the transformation of the mandate of Kosova Security Forces into the mandate of Armed Forces. Thus, the Emergency Management Agency, which is an executive agency within the MoI, has to undergo some reforms in order to fulfill its main role in managing incidents and natural disasters. In 2012 the Government approved the *Regulation on the methodology for risk assessment development* which derived from *the Law on Protection against Natural Disasters and other Disasters*. The purpose of this *Regulation* was to define the methodology in drafting risk assessments for the population, material goods and cultural goods from natural disasters and other disasters, stakeholders and participators in drafting the assessment, as well as procedures for its drafting and approval. The *Regulation* obliges that the assessments are to be carried out by legal entities whose activities are related to critical infrastructure facilities and legal persons according to *Directive 2003/105/EC amending Council Directive 96/82/EC on the control of major-accident hazards involving dangerous substances* and the so-called *SEVESO II – Directive* or *Directive 2012/18/EU*. This *Regulation*, however, was not implemented in practice and currently is being revised.

10.2. Implementation

The purpose of the *Law on Critical Infrastructure* is to *preserve and protect national and European critical infrastructure, protect citizens of the Republic of Kosova, prevent incidents and minimize potential damage to critical infrastructure, general wealth, economic and social losses, ensure government stability, and enhance resiliency*. The Republic of Kosova is at an early stage in developing and enforcing a CIP policy framework. Apart from the approval of the *Law*, the normative framework has yet to be completed as secondary legislation has not been drafted, the institutional mechanism has not been established, and the process of identification and designation of critical infrastructure has not started yet. Thus, there was little progress in implementing the provisions of the *Law*.

According to the *Law*, the MoI is the institution of reference for the implementation and monitoring of the progress of the Law. The following table provides an overview of the major roles and responsibilities of the relevant public institutions for the implementation of the Law:

Table 10.1 The Main Institutional Framework of the Republic of Kosova on CIP

MoI	*Government*	*Ministries*	*Parliament*
– Implement and supervise the Law **– Point of contact for national and ECI** **– Identify and prioritize the list of CI** **– Determine cross-cutting and sectoral criteria** **– Assign deputy security coordinators** **– Review OSPs** **– Engage in bilateral and/or multilateral discussion** **– Report to national parliament and European Commission** **– Initiate infringement procedures**	– Determine other sectors of CI – Assign sector coordinators – Designate national CIs – Review and revise the list of national CI – Designate ECIs – Provide financial incentives to owner/ operators of designed national and ECI	– Nominate sector coordinators – Support determination of cross-cutting and sectorial criteria – Engage in the review of OSPs	– Monitor the implementation of the Law, through parliamentary hearings and reports

The MoI with the support of TAIEX Expert Mission,[4] developed the *National Critical Infrastructure Plan 2019–2023* laying down a roadmap which consists of five phases as summarized below:

1. ***Administrative Arrangements for Initial Activities in Implementing the Law (January –July 2020)***
   During this phase, the MoI is expected to establish the relevant institutional mechanism and recruit the required civil servants with educational and professional background according to the sectors identified in the Law. In addition, the Critical Infrastructure Working Group (CIWG) will be established to serve as a national secretariat and composed of representatives of relevant

public institutions that have the mandate and authority over the sectors identified in the Law.

2. ***Identification and Designation of National and European Critical Infrastructure (July–December 2020)***

   In this phase, the identification and designation of national and ECI will begin, conducted with the support of international experts either through EU instruments or through other partners. The designation of ECIs, however, will depend on whether countries in the Western Balkans adopt the relevant legislation. An important activity during this phase is the start of a pilot project on developing OSPs in order to provide the owner/operators with the necessary OSP guidelines.
3. ***Operator Security Plan (January–December 2021)***

   The third phase will focus entirely on the OSPs and will cover activities such as the establishment of the registry of security coordinators and First Responders, development of procedures for the evaluation of OSP, and creation of an "early warning system".
4. ***National Annual Security Exercise and External Outreach (January–December 2022)***

   The national security exercise will be conducted as part of the National Response Plan with the inclusion of different CIP scenarios. The exercise would provide conclusions and recommendations to government authorities and CIP operators. Secondly, a regional forum would be organized to discuss regional cooperation on CIP-related trans-boundary externalities. Both activities will be submitted for support to the EU.
5. ***Review of the National Legislation and Framework on CIP (January–December 2023)***

In 2023 a thorough analysis and assessment of the legislative and policy framework on CIP will be conducted jointly by government officials and EU experts. The review will serve to improve the overall CIP framework, including the revision of the Law in order to address issues and/or further align them with the EU acquis.

In reference to the *Law on Critical Infrastructure*, the MoI is amending the *Law on Private Security Companies* adding a special type of private security service – Critical Infrastructure Security Services (CISS). The draft law has defined CISS as a *security service provided by a Private Security Company for securing critical infrastructure through authorized Security Workers, under the conditions and criteria set forth in this Law and relevant legislation*. The companies and security officers will be obliged to complete special training except for basic and other mandatory training. Moreover, the draft law stipulates that in order to obtain a license to provide Critical Infrastructure Security Services, the company shall have at least 20 basic Security Workers for critical infrastructure; a certified weapon storage room in accordance with the *Law on Weapons*; and a certified Control Centre.

10.3. Public-Private Partnership

The protection of critical infrastructure requires intensive cooperation and coordination between government authorities and owners/operators of critical infrastructure. This is of critical importance especially in countries that lack experience in implementing policy and legislation in CIP, as well as in countries that have recently adopted or are in the process of developing legislation. In the case of Kosova, the lack of experience and limited implementation capacities is also followed by the lack of well-defined and formal cooperation between the public and private sectors in CIP.

Thus, considering the importance of such cooperation the *Law on Critical Infrastructure* included a specific article – article 8 Public-Private Partnership. The article stipulates the following:

1. *National Critical Infrastructure protection requires coordination and cooperation with both public and private partners in order to ensure National Critical Infrastructure security and resiliency.*
2. *The Government of the Republic of Kosova, through MoI, will provide opportunities for engagement and critical infrastructure information sharing by organizing workshops, working groups, and meetings to strengthen public-private partnerships, and sharing relevant information for risks and threats to critical infrastructure, best practices and lessons learned by other countries.*
3. *The Government of the Republic of Kosova shall provide financial incentives for owners/operators of infrastructure designated as critical infrastructure.*

The inclusion of this article was done in order for the government to build and maintain open, trusted and continuous cooperation with the private sector. In addition, the government will support owners/operators of critical infrastructure to invest in security measures through financial incentives. These financial incentives are not agreed upon yet and they will be developed through a sublegal act.

In the cyber domain, the *National Cyber Security Strategy and Action Plan 2016–2019* have envisaged building public and private partnerships as a strategic objective. For example, the National Council for Cyber Security has included its structure the Kosova Banking Association while Internet Services Providers (ISPs) have not nominated a representative yet. The Regulatory Authority of Electronic and Postal Communications as well as the Police Sector for Investigations of Cyber Crime maintain regular meetings with ISPs. The new *National Cyber Security Strategy* will further focus on strengthening the public-private partnership through the establishment of formal structures for cooperation, joint activities for capacity building and joint projects.

10.4. Cooperation with European Commission

The Republic of Kosova and the European Union (EU) have signed the SAA on 27 October 2015, being the first contractual agreement between the two parties. The Republic of Kosova is a potential candidate country and has a membership perspective similar to the other countries of the Western Balkans. The European Commission cooperates with the Republic of Kosova bilaterally and in a regional context as Western Balkans.

From a bilateral perspective, the government and the EC maintain close cooperation and coordination through different structures and modalities. On the basis of the SAA, the Government has established the Subcommittee on Justice, Freedom and Security through which annual meetings are held between government institutions and the EC. These meetings serve to assess the alignment of Kosova policy and legislation with the EU acquis and monitor the implementation of Country Reports recommendations. The topic of CIP was only included in the 2019 agenda, however, it was discussed slightly. In addition, the Government prepares every year the *National Plan for Implementation of SAA* which sets up short-term and medium-term priorities for harmonization and alignment of national policy and legislation with EU acquis and policies, as well as activities to strengthen implementation capacities. Measures in relation to CIP and cyber security are mainly planned under chapter 24 Justice, Freedom and Security and but also on chapter 10 Information Society and Media. The EC organized in 2018 the Peer Review Expert Mission to assess compliance in the field of counter-terrorism and violent extremism. The scope of the mission included also the protection of critical infrastructure and the experts provided a number of recommendations. Another form of bilateral cooperation between the Republic of Kosova and the EU is through IPA funds. Until now there were no EU-funded projects in the protection of CI, however, the MoI has submitted the proposal to the EU for funding under the Instrument of Pre-Accession Assistance II (IPA) Funds for a project on Critical Infrastructure and Cyber Security; the project is expected to start in 2021/2022.

From a regional perspective, the MoI has participated in the meeting of the EPCIP Points of Contacts held in Romania in 2016 and Austria in 2018. These meetings provided excellent opportunities for non-EU countries to be informed about the implementation and developments at the EU level and in EU member states in protecting critical infrastructure. As the *Draft Law on Critical Infrastructure* was being developed, the presentations and discussions provided important information and helped draft the law. On October 2018 the Ministers of Interior of the Western Balkans and the European Commissioner for Migration, Home Affairs and Citizenship, representing the European Union, signed the *Joint Action Plan on Counter Terrorism for the Western Balkans*. Within the fifth objective *Strengthen the Protection of Citizens and Infrastructure*, each Western Balkan country is required to enhance the protection of its cyberspace in line with the *EU Cyber Security Strategy* and the *NIS Directive*, improve the protection of critical infrastructure, engage in developing preparedness and capacity to deal with CBRN risks and other measures. On the other hand, the EU as part of the external dimension of the EPCIP will invite Western Balkans countries to identify concrete projects and topics for cooperation on CIP. In that regard, the EC shall be engaged more proactively and regularly in regional initiatives in the Western Balkans. A regional project on the protection of critical infrastructure is urgently required to support countries in developing legislative and policy frameworks on CIP as well as to establish a framework of cooperation between countries in the Western Balkans.

10.5. Education and Training

Critical infrastructure protection is a relatively new field for government institutions and the private sector as well. The lack of expertise on CIP was evident while developing the *Concept Document on Identification and Protection of Critical Infrastructure* as well as the *Law on Critical Infrastructure,* where only a small portion of government institutions, namely Police and EM, had practical knowledge on CIP mainly due to their experience in conducting risks assessments. However, the support of international experts was required in order to develop the aforementioned documents.

In parallel to the development of the *Law on Critical Infrastructure*, efforts were made to ensure the support of the EU and other international organizations present in the Republic of Kosova to provide basic training and workshops for both public institutions and the private sector to better understand the CIP framework and prepare the human capacities to implement the *Law*. The only activity that was conducted in 2019 was the TAIEX Expert Mission which supported the development of the *National Critical Infrastructure Plan 2019–2023*. This comprehensive plan has identified the necessary training and exercises that are needed, and it also provides donors with the relevant information for potential programming of assistance.

The TAIEX instrument will continue to be used as a follow-up of the first expert mission held in October–November 2019 to develop secondary legislation, organize workshops, training and exercises and other activities to increase national and local capacities. As mentioned earlier, an EU-funded project on critical infrastructure and cyber security is expected to be launched once the pandemic crisis will be more under control and will provide comprehensive support to the MoI as the main beneficiary, but also to other relevant public institutions and the private sector.

While there is no specific program at any higher education institution in the field of critical infrastructure, there are plans to develop trainings modules for public safety agencies and owners/operators of critical infrastructures which will be offered by the Kosovo Academy for Public Safety (KAPS). KAPS is an internationally and domestically accredited institution that provides high-quality education, research and development for public safety agencies in the Republic of Kosovo. Each year the Academy trains thousands of public safety officials, by offering them basic, advanced and specialized courses in a range of areas. In the medium term,

KAPS plans to develop and implement curricula and modules for basic, advanced and specialized courses on critical infrastructure.

Chapter Conclusion

The Government of the Republic of Kosova has taken important steps in developing the normative and strategic framework on critical infrastructure protection and cyber security, which were inspired by EU legislation and policy as well as successful experiences of EU Member States.

The Republic of Kosova has approved the *Law on Critical Infrastructure* which was also inspired by the *Croatian Critical Infrastructure Act* which is a great reference document for other countries that are planning to develop a normative framework on CIP in line with EU acquis.

The approached used by the Republic of Kosova in transposing the *Council Directive 2008/114/EC*, especially in reference to ECIs is a good example as it provides a normative framework to engage in bilateral and/or multilateral activities to designate and protect critical infrastructures with regional impact. In addition, the *National Critical Infrastructure Plan 2019–2023*, can be considered by other countries as a roadmap for preparing thorough governance on critical infrastructure protection, including prioritization of activities and building of capacities to implement key aspects of National CIP Frameworks.

Provided the upcoming policy packages in the context of Network and Information Security and Resilience of Critical Entities, which aim at boosting further the cooperation among Member States of the EU, the European Commission shall launch regional projects to support countries in developing the necessary amendments to national frameworks and mechanism to establish cooperation platforms between countries in the region for protection of critical infrastructure with regional relevance, and also build the necessary human and technological capacities to improve and maintain CIP and cybersecurity frameworks.

Notes

1. Concept Document is a policy paper similar to Regulatory Impact Assessment and it is mandatory to be approved by the Government before any new legislative act, supplement and change of legislation or policy with significant implications.
2. Counter Terrorism Strategy and Action Plans 2009–2012, 2012–2017 and 2018–2023.
3. The Strategy was approved by the National Security Council in 2019 but due to early national elections in October 2019, the government was not able to proceed with the strategy to the Parliament.
4. The mission was held from 29 October to 1 November 2919.

EASTERN PARTNERSHIP

3

Ukraine
Belarus
Moldova
Georgia
Armenia
Azerbaijan

DOI: 10.4324/9781003273769-13

Chapter 11

Ukraine

Oleksandr Sukhodolia

Contents

The country's capacity to confront internal and external threats and respond adequately to challenges depends on many factors to which the developed states assign, inter alia, the levels of CIPR.

For Ukraine, the urgency of critical infrastructure protection (hereafter CIP) was actualized not only by general world trends toward increasing natural and man-made disasters, terrorist threats, sophisticated cyberattacks against infrastructure but also by Russian hybrid aggression in 2014. The hybrid war (Horbulin, 2017) emphasized the utmost importance to protect systems, facilities, and resources critically important for the functioning of the society, social and economic development, and continuity of governance. It underlined the need to establish a nation-wide systematic "all-hazard approach" to CIP, considering the mutual interface between critical infrastructures with a focus on securing continuity of vital for social services.

However, the state/national protection systems, as well as crisis response systems available in Ukraine,[1] are focused on their "own" threats and risks, and, therefore,

DOI: 10.4324/9781003273769-14

to a large extent autonomous (Sukhodolia, 2017: 89–107; 123–126). The tasks of enhancing CIP resilience differs from the priorities of the existing systems (civil protection, counter-terrorism, physical protection, cyber threat counteraction, etc.) and demand legal, institutional, and organizational novelties.

The wider discussion of ways to adapt the national security system to modern threats was formalized through development by the NIS, the *Green Paper on Critical Infrastructure Protection in 2014–2015* (Biriukov, 2015).

11.1. Introduction

Ukraine clearly stated its Euro-Atlantic aspirations and wish to join both NATO and the European Union and in 2014 by signing the Association Agreement between the European Union and Ukraine (European Atomic Energy Community, European Union, The Member States, Ukraine, 2014). Ukraine believes that the harmonization of approaches for CIP with those actively implemented by the EU and NATO will facilitate the improvement of the national security and resilience of Ukraine as well as the whole of Europe.

Though Association Agreement does not establish a formal obligation of Ukraine on CIP the EU *Council Directive 2008/114/EC* and *Directive (EU) 1148/2016* became orientation for Ukraine's internal efforts in developing a legislative basis for a CIP policy (Council of the European Union, 2008; European Parliament and of the Council of the European Union, 2016).

In 2015, the revised *National Security Strategy for Ukraine* (President of Ukraine, 2015) for the first time singles out threats to critical infrastructure as well as threats to cybersecurity. The security of critical infrastructure has been mentioned as one of the key areas of the state policy for national security, and its priorities have been identified. In 2016, the Decree of the President of Ukraine had approved the *Cyber Security Strategy of Ukraine* (President of Ukraine, 2016a). These first acts of legislation are in full compliance with the priorities mentioned in EU Directives.

In 2016–2017, according to the *Decision of the National Security and Defense Council of Ukraine "On improvement of the measures to ensure the protection of critical infrastructure objects"* (President of Ukraine, 2016b), there were measures taken for ensuring CIP as well as improving the legal basis for the establishment of a comprehensive state system on CIP in Ukraine.

The *Law of Ukraine "On the basic principles of ensuring the cybersecurity of Ukraine"* (Parliament of Ukraine, 2017) had determined the legal and organizational basis of ensuring the protection of the vital interests of citizens, society and state, and national interests of Ukraine in cyberspace. The Law fixed the main objectives, directions, and principles of state policy in the field of cybersecurity as well.

The "*Concept for building a state critical infrastructure protection system in Ukraine*" (Cabinet of Ministers of Ukraine, 2017) approved by the Ukrainian government had outlined priorities for the establishment of a CIP state system and determined main directions, mechanisms, and terms for comprehensive legislative regulation of CIP activities (Sukhodolia, 2017: 56–68).

The goals of the concept have to be achieved through several measures:

- development of *the Law of Ukraine* on CIP
- working out a unified methodological framework for activities within the CIP system, identification, and designation of CI
- development of regulations on CI
- establishment of a public-private partnership in the field
- development of information-exchange procedures for risk assessment and sharing best practices
- development of national CIP programs/plans
- further harmonization of Ukrainian and EU legislation in the field of CIP

The Cabinet Ministers of Ukraine submitted the draft of *Law of Ukraine "On critical infrastructure and its protection"* to the Parliament of Ukraine on May 27, 2019 (Cabinet of Ministers of Ukraine, 2019a). The proposed draft of law establishes the CIP state system (as a set of organizational, legislative, scientific, and other measures aimed to ensure CI security and resilience), defines its fundamental principles of operation, application of "*all-hazard approach*" and common approaches to the management system at the state and local levels, and identifies principles of interaction between state authorities, private businesses, society, and public (Sukhodolia, 2019: 200–219).

The draft law describes the CI as a set of objects that are strategically important for the economy and national security, whose malfunctioning may be detrimental to vital national interests. This definition integrates an objective and functional approach to interpreting CI and reflects the definition given in EU regulations (*Council Directive 2008/114/EC*). Namely, the CI is a set of objects (material resources, fixed assets), systems, or their parts, which are essential for the maintenance of vital functions of society and whose disruption of functioning will have a significant impact on the country and will lead to the inability to provide these functions. The draft law indicates that critical infrastructure consists of sectors that, in turn, contain critical infrastructure objects.

Summarizing, it has to be stressed that Ukraine's effort in improving CIP is inspired by and is correlated with approaches reflected in the EU legislation. The goal of the CIP policy in Ukraine is to secure the supply of vital goods and services to the population, society, business, and state. The main priorities of the CIP system are to warrant the uninterrupted and sustainable operation of critical infrastructure assets in prescribed modes of operation and to be able to prevent destruction or irreparable harm, stoppage, or loss of control of critical infrastructure assets as a result of threats' impact, as well as to assure quick recovery of their operation where it was disrupted.

11.2. Implementation

At the time of assessment, Ukraine still has not had implemented the separate CIP state management system. Nevertheless, the main concept of the system was agreed upon (Sukhodolia, 2017, 2019). The organizational model in Ukraine is developed as well as a set of principles, mechanisms, and tools that can be used.

Defining principles of the CIP system functioning:

- introducing common critical infrastructure sector-based working definitions and terminology
- organizing interaction between involved entities of the system
- introducing a unified methodology for a risk assessment under the all-hazard approach
- determining a methodology for assigning infrastructure facilities to the National Critical Infrastructure List
- establishing requirements for planning measures in the field of critical infrastructure (contingency plans, emergency plans, training (exercise) plans)
- determining powers, tasks, and responsibilities of state agencies and authorities as well as duties, and responsibilities of owners (operators) of critical infrastructure facilities
- determining the legislative basis for public-private partnership (duties and responsibilities, contact persons, interaction, information exchange) and support of incentives to invest in CIP

Defining an organizational structure of the system:

- establishing an authority responsible for the coordination of activities of all involved entities
- determining tasks and responsibilities of security and defense sector entities (law enforcement, intelligence and counterintelligence, counter-terrorism and cyber protection, etc.)
- determining the tasks and responsibilities of state agencies (ministries) responsible for CIP in different sectors
- determining the duties and responsibilities of operators of critical infrastructure facilities

The *Law of Ukraine "On the basic principles of ensuring the cybersecurity of Ukraine"* had determined the main subjects within the national cybersecurity system, namely: State Service of Special Communication and Information Protection, National Police, Security Service of Ukraine, Ministry of Defense, General Staff of the Armed Forces, intelligence agencies, and the National Bank.

In 2019, the Ukrainian government approved basic requirements for critical infrastructure cybersecurity (Cabinet of Ministers of Ukraine, 2019b)

Figure 11.1 Description of proposed state system on critical infrastructure protection in Ukraine

requesting from critical infrastructure operators to create a comprehensive system of cyber protection of the information infrastructure of the critical infrastructure facilities.

The draft of *Law of Ukraine "On Critical Infrastructure and its Protection"* gives additional tasks to the subjects within the national cybersecurity system and determines powers, tasks, and responsibilities of subjects responsible for CIP in different critical infrastructure sectors.

The main subjects within the CIP system, as proposed (Sukhodolia, 2019: 87–97; Cabinet of Ministers of Ukraine, 2019a), are Government Center on Critical Infrastructure Protection, sectoral authorities responsible for the protection of specific sector critical infrastructure (10 sectors),[2] functional agencies responsible for security and response to specific threats (six existing security and response systems),[3] local authorities, and operators of critical infrastructure.

The general description of the proposed CIP state system (Figure 11.1) is as follows:

- The Government of Ukraine (the Cabinet of Ministers of Ukraine) establishes a separate authority responsible for coordination CIP (the Governmental Center on Critical Infrastructure Protection)

- The Governmental Center with support from the sectoral ministries and functional agencies, according to a special procedure, designs the National Critical Infrastructure List
- The Governmental Center with support from sectoral ministries and functional agencies conducts yearly National Risks Assessments prioritizing actions of all involved entities
- The Government of Ukraine approves the National Plan for Critical Information Protection for coordination of activity of all involved entities at the national level
- Sectoral ministries develop and implement sectoral plans that reflect the specificity of critical infrastructure sectors
- Operators with support of sectoral ministries and functional agencies, according to a special procedure, develop the OSP.

As sectoral authorities, it is proposed to nominate some ministries and agencies that, by the decision of the government, will take responsibility and implement the state CIP policy within certain CI sectors (sectors – a set of CI objects that belong to one sector of the economy and/or have a common functional orientation).

It was proposed to nominate sectoral authorities upon their responsibility for selected vital services/functions (Table 11.1) (Sukhodolia, O. et al., 2019: 174–176; Potii O. et al. 2020).

As sectoral authorities, nine ministries and agencies were nominated, which by the decision of the government (Cabinet of Ministers of Ukraine, 2020b) will take responsibility to implement state CIP policies within their sectors. Depending upon a structure of the government,[4] it would be the Ministry of Energy, the Ministry of Infrastructure, the Ministry of Health, the National Bank, the Ministry of Regional Development, the Ministry of Economy, the Ministry of Internal Affairs, and others.

The following support from the Sectoral ministries and Functional agencies support from the Sectoral ministries and Functional agencies functional authorities are proposed to nominate the next six institutions: Security Service of Ukraine, State Emergency Service, State Nuclear Regulatory Inspectorate, State Service of Special Communication and Information Protection, National Police and Guard of Ukraine, and General Staff of the Armed Forces.

The National Critical Infrastructure List, to determine the required level of protection of CI facilities as well as tasks of responsible entities, is categorized as follows:

- *critically vital facilities* – facilities of nationwide and cross-border importance with complex ties and significant impact on other infrastructure
- *vital facilities* – malfunction of such facilities will lead to a regional level crisis
- *important facilities* – facilities that are important for the continuity of vital social functions and services (at operator's domain)

Table 11.1 List of Critical Infrastructure Sectors of Ukraine and Relevant Vital Services (Functions)

№	*Responsible Authority*	*Sector*	*Subsector*	*Vital Service (Function) Type*
1	Ministry of Energy	Fuel and energy sector	Electricity	Electricity production and supply
				Transmission systems and management
				Electricity distribution system services
			Oil and petroleum products	Oil production
				Transmission of oil and petroleum products
				Oil purification, refining, and treatment
			Gas	Gas production
				Gas transmission system operation
				Gas distribution operation
				Storage of natural gas
			Nuclear power	Production of electricity at nuclear power plants
				Nuclear power plant operation
				Nuclear fuel treatment
2	Ministry of Digitalization	Information sector	Information technologies	Data processing services in data centers (DCs) and/or cloud storage
				Web services and electronic trust services
				Distribution of television (including digital) and radio signals

(Continued)

Table 11.1 *Continued*

№	*Responsible Authority*	*Sector*	*Subsector*	*Vital Service (Function) Type*
			Telecommunications	Internet Domain Name System (DNS) support
				Transmission of data through fixed, mobile, and special communications, and telecommunication networks
3	Ministry of Regional Development	Life support networks	Drinking water	Drinking water supply and treatment
			Waste and sewage	Waste treatment
				Centralized drainage systems operation
			Heat supply	Heat supply
4	Ministry of Economy	Food industry and agriculture		Production and processing of agricultural and/or food products
				Operation of large irrigation systems and canals
				Strategic state stocks and reserves operation
5	Ministry of Health	Health care		Emergency medical care provision
				Hospital and outpatient treatment and other medical services provision
				Production, procurement, and supply of medicines, vaccines, blood, and other medicines, and medical equipment
				eHealth services provision
				Control of infectious diseases and/or epidemics

6	The National Bank	Financial sector		Banking and insurance services
				Stock exchange services and stock market
7	Ministry of Infrastructure	Transportation and mail	Air transport	Air traffic control
				Air transportation (air transport work)
				Airports and ancillary facilities operating at airports
			Road transport	Bus transportation services (long distance, international, and tourist transportation)
				City transportation services (buses, trams, trolleybuses, subway)
				Maintenance of transport infrastructure (roads, bridges, tunnels, overpasses)
				Traffic control services and traffic management
			Railway transport	Passenger rail transportation services
				Freight rail transportation
				Operation and maintenance of the railway
				Station work
			Sea and river transport	Control and management of shipping
				Inland, sea, or coastal passenger and freight operations
				Operation and maintenance of infrastructure (channels, dams, fairways, etc.)
			Mail	Mail services

(Continued)

Table 11.1 *Continued*

№	*Responsible Authority*	*Sector*	*Subsector*	*Vital Service (Function) Type*
8	Ministry of Internal Affairs	Public security and the rule of law		Ensuring public order and security
				Judicial and criminal systems functioning
9	Ministry of Economy	Industry	Chemical and nuclear industry	Production of industrial gas
				Manufacture of fertilizers or nitrogen compounds
				Production of pesticides or other agrochemical products
				Production of explosives
				Manufacture of other basic inorganic substances
				Manufacture of other basic organic chemicals
			The military industry	Manufacture and supply of weapons and military equipment (military-industrial complex)
			Space and aviation	Manufacturing and supply of rocket and aerospace industry products

10	Ministry of Internal Affairs	Civil protection of the population and territories		Emergency notification of the population, liquidation and disaster recovery, organization and coordination of rescue of residents and property
11	Ministry of Environment Protection	Ecological safety		Air pollution monitoring and early warning
				Meteorological observation and early warning
				Surface water monitoring (rivers, lakes) and early warning
				Monitoring and control of marine pollution

Source: Author's interpretation

Critically vital facilities are designed to include facilities identified as the ECI. Vital and important facilities will relate to the National Critical Infrastructure.

In Ukraine, the set of main tools of a CIP system are already partly designed and developed. The following tools propose to be implemented within the CIP system: regulation (standards, instructions, design basis threat), planning (national plan, sectoral plans, OSP), collaboration (information exchange and interaction procedures, communication systems, emergency commissions), and training (education, training, exercises).

All tools are supposed to comprise an identification of important assets, a risk assessment, an identification, selection and prioritization of countermeasures, and a restoration of critical infrastructure functioning.

Currently, Ukraine is actively working on establishing the CIP system. Sectoral ministries have established interagency groups for the identification and designation of critical infrastructure as well as the development of requirements for CIP within their fields of responsibilities. Functional agencies are involved in the CIP process through supporting actions of sectoral ministries and strengthening the protection of critical infrastructure facilities (counterintelligence, physical, and cybersecurity measures) according to the orders of the Government of Ukraine.

The Security Service of Ukraine took a leading role in CIP concept implementation following priorities of reform of the Security and Defense Sector of Ukraine (President of Ukraine, 2016b). The Security Service of Ukraine is tasked to provide threat identification and counterintelligence protection critical infrastructure as well as provide informational exchange and support to efforts of other ministries and government agencies.

State Service of Special Communication and Information Protection of Ukraine is playing a significant role in developing the CI system. Due to its efforts, there were developed basic requirements to the critical infrastructure operators regarding the cyber protection measures of the information infrastructure (Cabinet of Ministers of Ukraine, 2019b) and the methodology for designation of the CII List (Cabinet of Ministers of Ukraine, 2020a). The methodology of conducting independent auditing of the cybersecurity of the critical infrastructure objects is under consideration (State Service of Special Communication and Information Protection, 2020).

11.3. Public-Private Partnership

In Ukraine, as in the rest of European countries, it becomes obvious that critical infrastructure assets are or soon will be mostly privately owned. Thus, effective public-private partnership (hereafter PPP) becomes a key element of the successful and sustainable CIP policy.

At the same time, the concept of PPP is still a new one for Ukraine. Though the *Law of Ukraine "On Public-Private Partnership"* introduced the PPP mechanisms in

2010 they were targeted to attract private investment (mainly construction projects, concessions) and did not refer to CIP investments.

The current PPP on CIP issues is restricted only to established working groups in sectoral ministries to identify critical infrastructure facilities and potential threats, to suggest improved legislation on generic issues. Public private partnership in the security field is rather rare, due to the restrictive state regulation regarding classified information sharing and involvement of non-government actors into the activity of the defense and security sector. On the other hand, there is a reluctance of the private sector to grant access to state authorities for commercially sensitive information.

A more developed form of cooperation between government and non-government subjects exists in the field of cybersecurity. The legislation in the field requires the private sector to implement protection measures and cooperate with state agencies in the field. The concept of PPP is aimed at creating a system for the timely detection and neutralization of cyber threats, increasing the digital literacy of citizens, exchanging information between state agencies and the private sector, partnership and coordinating with teams responding to computer emergency, attracting expert potential, educational institutions, scientific and research organizations in development and implementation of cybersecurity measures, introducing a mechanism for public control of the effectiveness of cybersecurity measures, creating a system for training special forces, and so on. Within the sector a CERT-UA team is established through which many private actors could come to help operators of critical infrastructure to cope with cyber threats.

But in general, Ukraine still needs a proper regulatory basis for PPP in CIP with the description of mutual obligations of the private and state agencies concerning CIP. There have to be developed the procedures for the involvement of the public and commercial sector in risk analysis and contingency planning, as well as established coordination and responsibility-sharing mechanisms (including financial responsibilities) mechanisms between government and non-government subjects.

11.4. Cooperation with European Commission

Ukraine does not have a well-established form of cooperation with the EU in the CIP field. It is because CIP was not included in the priorities of EU–Ukraine cooperation earlier, and Ukraine still needs legislation that will initiate such cooperation. Nevertheless, due to its geographical location, Ukraine has especially tight links with the energy and transport infrastructure of the EU Member States. The EU and Ukraine are aware of the importance of security and information and communication technologies as both suffered the NotPetya virus, the destructive impact of which had been felt in Ukraine and a number of countries in Europe. By signing the Association Agreement, Ukraine took a formal obligation to implement EU legislation as well.

At present, Ukraine is engaged in close cooperation with the EU on different aspects of CIP. There are well-developed forms of cooperation on securing steady functioning of the Ukrainian gas transportation system, which could be designated as ECI s the system's disruption or destruction would have a significant impact on at least two Member States (*Council Directive 2008/114/EC*), though not formally belonging to the EU country.

The Association Agreement has made Ukraine assure compliance with EU energy legislation such as *Regulation (EU) 2017/1938 concerning measures to safeguard the security of gas supply* (European Parliament and of the Council of the European Union, 2017). Ukraine approved the Preventive Action Plans on the security of gas supply (MECI, 2015a) as well as electricity supply (MECI, 2018) and Emergency Plan (MECI, 2015b). Under the EU energy legislation framework, an "early warning mechanism" dealing with transboundary CEIs has been established by the EU and Ukraine as well (Cabinet of Ministers of Ukraine, 2019c). Ukraine develops a sectoral cybersecurity system according to the latest EU regulations and recommendations on cybersecurity measures in the energy sector as well.

The cybersecurity, development of capabilities in combating cyber and hybrid threats is at the center of attention of EU–Ukraine cooperation. In 2017–2018, the EU implemented a number of activities under the Technical Assistance and Information Exchange (TAIEX) program in three areas: the creation of an appropriate legislative framework in Ukraine, the creation of PPP and promotion of organizational aspects of national cybersecurity structures, and support of technical abilities and skills in state authorities responsible for cybersecurity. The EU assists Ukraine through its EU Advisory Mission to Ukraine (EUAM). More than 2.5 million euros were allocated by the EUAM to various projects helping Ukraine in the area of cybersecurity. The mission promotes the improvement of technical equipment of Ukrainian law enforcement agencies, conducts training, and exchanges experience and panel discussions.

Ukraine is interested in engaging in the work of the European Union Agency for Network and Information Security, the activities of the European Cybersecurity Industrial, Technology and Research Competence Center, as well as training of the EU on coordination mechanisms for the joint response of the EU and the Member States for large-scale incidents and crisis situations in the field of cybersecurity.

Ukraine has close relations with the EU on the issues of prevention and response to natural and technogenic threats and catastrophes. Recently, EU–Ukraine cooperation in the field started to focus on supporting Ukraine in the development of the country's system for protecting critical infrastructure and chemical security. The EU-funded program on prevention, preparedness, and response to natural and man-made disasters in the EaP countries (PPRD East 2) started to support the State Emergency Service of Ukraine in its efforts.

11.5. Education and Training

Ukraine has a well-developed system of education and training in the field of the response to natural and technogenic threats, which includes the issue of critical facilities protection. There are a number of universities that propose programs in the field as well as separate research institutes that support the practical activity of the government.

In the last decade, Ukraine established a number of educational programs in the field of information security and information security management. A number of departments on cyber security and information protection have been established in different universities. Currently, they provide not only education services but also research and training in the field of cybersecurity.

The issue of CIP is that it still is not implemented in a formal educational process though the preparations are underway. In Ukraine, since 2015, a series of events has been organized to support the implementation of the CIP system. A series of workshops and seminars discussing the problem with public servants and experts has been conducted by the National Institute for Strategic Studies of Ukraine. Expert meetings and conferences discussing the CIP were held by Ukrainian ministries and governmental agencies. Ukrainian efforts in raising awareness and education of staff of some ministries have been supported through programs of international cooperation.

The following are some examples of such activity (Sukhodolia, 2017: 15–116):

- Seminar "Establishing a system of personnel training in Ukraine in the field of CIP" (9 November 2016), conducted by the NISS together with NATO Professional Development Program in Ukraine
- Workshop "The main approaches to planning of actions in the event of a crisis on objects of critical energy infrastructure" (13 April 2017), conducted by the NISS
- Expert meeting "Best practice of public relations in the event of serious incidents on critical infrastructure objects" (11 May 2017), conducted by the NISS
- Workshops "Critical National Infrastructure" (19–23 June 2017) and "Critical National Infrastructure: Energy Sector" (4–7 September 2017) was supported by the UK Cabinet Office Emergency Planning College

In October 2017, Ukraine held the first national-level Table-Top Exercise (TTX) on the issue of resilience of critical infrastructure. The TTX "Coherent Resilience 2017" (NISS, 2017) was organized by the NISS, the NATO Energy Security Centre of Excellence, and the Ministry of Energy and Coal Industry of Ukraine and became the practical step in NATO support of Ukraine in reforms of Ukrainian national security sector. The TTX had goals to check existing procedures on prevention, protection, and response on incidents related to CEI and to facilitate the cooperation of different agencies in their action to provide the resilience of the National Power System (Sukhodolia, 2018).

The target audience of TTX was personnel of government agencies and ministries in the field of energy, emergency service, and national security that included energy government officials, military personnel, national police, and other institutions and agencies responsible for the protection and building of resilience of electricity supply by improving plans, procedures, and processes at a national level.

In October 2018, a similar TTX was held with a focus on regional critical infrastructure. The TTX CORE2018 with focus on sectors of financial and transport infrastructure was organized by the MIA of Ukraine and the UK Cabinet Office Emergency Planning College.

The CIP issues were introduced in a curriculum of the universities that provide educational and training programs for the security and defense sector of Ukraine. Depending upon specificities, CIP modules are included in counterintelligence, law enforcement, emergency, and cybersecurity programs. The CIP module became a part of the Strategic Leadership Program initiated by the Kyiv-Mohyla Academy Business School in 2017, which was designed to prepare strategic-level personnel from the military and security sector, public service, NGOs, and private sectors.

Chapter Conclusion

Despite tangible results being achieved, there is still a lot of work ahead to build an effective CIP system in Ukraine. Future efforts have to focus on establishing a framework of robust interaction, adequate levels of cooperation and interaction among all involved stakeholders, well-developed and sustainable PPP, adequate training and education capabilities, and involvement in international cooperation in this field.

The specification of the CIP system's tasks requires the development and implementation of legislation. Ukraine needs the approval of the relevant *Law of Ukraine "On critical infrastructure and its protection"* by the Parliament as soon as possible. At the same time, the Government of Ukraine has to take the leading role in establishing the system and without delay establish separate authorities responsible for coordinating CIP and approving the National Critical Infrastructure List.

It is very beneficial for Ukraine to have at hand experience of EU countries to avoid mistakes of policy implementation. It is important not only through using "best practice" methodology or legislation but also through direct involvement of experts in development pieces of legislation.

Notes

1. The most important are four systems: the Unified State System for Civil Protection, the Unified State System for Prevention of Responding to and Suppressing Terrorist Acts and Mitigating Their Consequences, the State Physical Protection System, the National Cyber Security System.

2. Authorities responsible for energy (including nuclear), transport (land, air, and water), communication, health, finance, water and heat supply (for households), food, defense, governance, emergency.
3. Authorities responsible for physical protection of facilities (the State Physical Protection System), civil protection (the Unified State System for Civil Protection), cybersecurity (the National Cyber Security System), counter-terrorism system (the Unified State System for Prevention of, Responding to and Suppressing Terrorist Acts and Mitigating Their Consequences), territorial defense (the System Territorial Defense), and law and order protection (the Law Enforcement System).
4. The list of responsible ministries is changing according to changes in governmental structure and responsibilities of ministries.

Chapter 12

Belarus

Nadya Karpenko

Contents

The purpose of the present chapter is to assess the state of CIPR in Belarus from a general point of view and with the respect to the EU neighborhood policy. The contribution attempts to place Belarus into the regional context and provide a snapshot of how CI is conceptualized and framed in the national legal discourse. It also offers an overview of the most important policy and normative acts that contain references to CI with a view to identifying strengths and shortcomings of the existing legal framework.

12.1. Introduction

Just like any other state today, Belarus is exposed to a wide range of modern threats and risks to its infrastructure, critical facilities, networks and information systems.

DOI: 10.4324/9781003273769-15

Its energy and transport infrastructures are tightly linked to the external market leading to a high degree of interconnection and dependency upon global CIP trends. Belarus is one of the key and the shortest transit states for European gas supply coming from Russia. It is also home to two of the largest oil refineries in the post-Soviet space and the world's longest oil pipeline network – the Druzhba pipeline (Balmaceda, 2006).

In spite of this, however, Belarus considerably lags behind the EU in the sphere of CIP. In the first place, its domestic legislation does not seem to categorize important assets and facilities as CI. The infrastructures that count as critical in accordance with the internationally agreed understanding of CI remain underconceptualized and regulated by fragmented and oftentimes outdated legal acts (Perevalov, 2016). Up until today, Belarus has not implemented any specific, consolidated critical infrastructure protection law and no such adoption is included in the Legislative Plan for 2020.

This situation emerges despite the fact that CI is referenced in several legal documents dating back to as early as 2010. Namely, the *National Security Concept of the Republic of Belarus* (President of Belarus, 2010) mentions CI and "critical informatization objects" under four different articles. In accordance with article 14, "safe and sustainable functioning of the critical informatization objects" belongs to the national interests of the Republic of Belarus in the information sphere. Further on, articles 27 and 34 stipulate that "the disruption in the functioning of critical informatization objects" constitutes one of the potential or existing threats to the national security of the Republic of Belarus; and identify "the shortcomings of the critical informatization objects protection" among the source of internal national security threats. Finally, article 51 of the Concept identifies the need to strengthen the material and technical base for the protection of critical objects, critical infrastructures and critical systems (President of Belarus, 2010).

A loose definition of CI is also found in the *Presidential Decree of October 25, 2011, No. 486 "On some measures for safety of critical informatization objects"*. The document gives definitions to "critical informatization object" and critical infrastructure:

- Critical informatization object means informatization object which

  1. serves to maintain ecologically dangerous or socially significant production facilities, and the disruption of which can lead to an emergency situation; 2. performs functions of information system, the disruption of which can have a considerable negative impact on national security in political, economic, social, information, ecological, and other dimensions; 3. provides information services, a partial or a full termination or breakdown of which can have a considerable negative impact on national security in political, economic, social, information, ecological, and other dimensions.

- Critical infrastructure is "infrastructure which is essential for the state, and the disruption or destruction of which would have a significant negative impact on national security" (President of Belarus, 2011).

Despite this rather early attempt to give a legal definition to CI, the *National Security Concept and the Decree "On some measures for safety of critical informatization objects"* are among the very few legal documents in Belarus that reference CI as a distinct and separate category. To date, the concept has remained largely elusive and is usually subsumed within the catch-all categories of "critical objects" or "important state facilities".

The shortcomings of the conceptual framework and classification scheme are pointed out in several scientific publications. According to Perevalov (2016) and Yurtsov (2016), Belarus lacks a single national legislative act governing CIP in Belarus and a strategic document establishing CIP policy is long overdue. Certain CI and critical objects related matters and provisions can be found in several legal documents and regulatory acts, as well as in separate resolutions of the President or respective ministries:

- *The Presidential decree of August 17, 2015, No. 356 "About the commission on counteraction to extremism and fight against terrorism"*
- The *Law of the Republic of Belarus "On the Protection of the Population and Territories from Natural and Technogenic Emergency Situations"* (Law No. 141-Z of May 5, 1998)
- The *Law of The Republic of Belarus "On the Fight Against Terrorism"* (Law No. 77–3 of January 4, 2002)
- The *Law of The Republic of Belarus "On industrial safety of hazardous production facilities"* (Law No. 363–3 of January 10, 2000)
- The *Law of The Republic of Belarus "On industrial safety"* (Law No. 354-Z of January 5, 2016)
- The *Resolution of the Ministry of Architecture of the Republic of Belarus of July 20, 2011, No. 40 "On the approval of the list of critical objects that require protection against terrorist attacks"*

Most of these legal acts identify potential vulnerabilities of various infrastructures belonging to the energy or transport sector, as well as industrial systems, and production assets. However, due to the absence of a conceptual framework, these assets are not designated as belonging to CI.

Instead, for instance, according to the *Resolution of the Ministry of Architecture of the Republic of Belarus of July 20, 15, 2011, No. 40 "On the approval of the list of critical objects that require protection against terrorist attacks"*, the following objects are designated as "requiring special protection": nuclear facilities and waste depositories, chemical and petrochemical infrastructure, hydraulic facilities and installations, fuel depots, oil terminals, high-power stations, marine facilities, underground railway systems, airports with the total runway length exceeding 1,800 meters, railway junctions and hub station, electric communication and television network infrastructure, public facilities and so on (Ministry of Architecture, 2011).

The *Law of The Republic of Belarus "On industrial safety"* further extends the list of "potentially hazardous objects and infrastructures" to include different types of gas facilities, cross-country, long-distance and short-distance gas pipelines, oil and oil products pipeline, gas measuring, gas distribution and gas delivering stations. The Law states that accidents on oil, and gas pipelines can have a significant negative impact on economic and human safety (House of Representatives of Belarus, 2016).

The Presidential Decree of August 17, 2015, No. 356 "About the commission on counteraction to extremism and fight against terrorism" defines Critical Objects (CO) as

> objects of social, production, engineering and transport, energy, information and communication and other infrastructures the disruption of which through the act of terrorism or civil unrest can lead to destabilization of public order and advancement of terrorist goals and (or) entail human victims, damage to human health or environment, significant material damage and disruption of daily life of people.
>
> (President of Belarus, 2015)

Protecting CO and CI from terrorists received increasing attention in Belarus in the 2000s. The growing importance attributed to energy and transport infrastructure protection is reflected in the amendments to The *Law of The Republic of Belarus "On the Fight Against Terrorism"* (House of Representatives of Belarus, 2002). The Law was amended twice since its adoption. In 2012, article 3 paragraph 10 was revised and the list of potential targets of terrorist attacks was extended to include different types of transmission pipeline transport and facilities. In 2018, the Law was supplemented with a paragraph in article 3 on "The prevention of terrorist attacks against critical objects". Pursuant to the new paragraph, state and non-state operators of CO are required to carry out the necessary protection measures, security activities and other measures to prevent terrorist activity and to minimize and mitigate the effects of any such occurrences.

Belarus' newest *National Concept of Energy Security Till 2035* is primarily concerned with the political and economic dimensions of energy security (energy independence and diversification of fuels) and not the safety of energy infrastructure. Nevertheless, the definition of Energy Security specifies accidents, natural disasters, and terrorism among the major potential threats to the domestic energy sector (Government of Belarus, 2015a).

12.1.1 Information and Communication Technology Sector

Most recently, Belarus has been focused on adopting progressive legislation in the field of ICT sector security. Cyberspace in Belarus is tightly controlled and the security of network and data protection appear to be assigned the highest policy priority.

The *Strategy of Informatization Development in Belarus for the period 2016–2022* lists the major priorities for ICT development, including infrastructure connectivity, information and communication infrastructure protection, and digital infrastructure security (Government of Belarus, 2015b).

At the moment, the main laws governing information and data protection in Belarus are the *Law "On Information, Informatization and Data Protection" No. 455-Z* (House of Representatives of Belarus, 2008a) and the *Law on Population Register No. 418-Z* (House of Representatives of Belarus, 2008b). The Legislative Plan for 2020 provides for a draft of a new version of the *Law "On Information, Informatization and Data Protection"* (The Official internet Portal of the President of the Republic of Belarus, 2020) which is expected to come into force in 2021. Among other things, this Law will introduce new rules on personal data processing in Belarus.

Presidential Decree No. 449 "On the enhancement of state regulation in the field of information security" elaborates the classification scheme for critical informatization objects and established the following criteria for assessing the criticality of an object: criteria of social significance, economic significance, environmental and informational significance (President of Belarus, 2019).

Another milestone document in the cyber domain is the *Information Security Concept of the Republic of Belarus* (Security Council of Belarus, 2019). Although it makes no references to any of the EU Directives, the concept is said to "take into account the main provisions of acts of international organizations, including resolutions of the UN General Assembly and OSCE recommendations" (The Official Internet Portal of the President of the Republic of Belarus, 2019).

12.1.2 Sources of Legislative Inspiration

In 2004 Belarus adopted the *Law "On Technical Regulation and Standardization"* which addressed the matter of pipeline safety. The Law provided for the creation of a series of technical regulatory normative acts governing pipeline safety and took direct inspiration in a respective EU Directive and EU best practices (Zoï Environment Network, 2015).

With the exceptions of the aforementioned legal act, no evidence has been found of direct legislative borrowing from the EPCIP policy package, the Cybersecurity policy package or any other CIP-related EU legal acts.

Instead, two other external sources of inspiration in legislative drafting can be named: the Commonwealth of Independent States Inter-Parliamentary Assembly (CIS-IPA) and the Parliamentary Assembly of the Collective Security Treaty Organisation (PA CSTO). Belarus is a current member of both organizations. Since their establishment, CIS-IPA and PA CSTO developed and approved several Model Laws and recommendations, which clearly reflect the high priority attributed to the matters of critical facilities and information systems security in the regional security agenda. Among them are:

- the CIS Model *Law of 28 November 2014 No. 41–14 "On the critical objects of information and communication infrastructure"*
- *"Recommendations on harmonization of national legislation of the CSTO member states on critical infrastructure protection"* (Parliamentary Assembly of the Collective Security Treaty Organization, 2014)
- *"Recommendations on improving the legislative support for countering technological terrorism in the CSTO member states"* (Parliamentary Assembly of the Collective Security Treaty Organisation, 2016)
- CSTO Model *Law "On the safety of critical facilitie*s" (2018)
- CSTO Model *Law "On Countering Technological Terrorism at the Fuel and Energy Complex"* (still in development)

The documents listed earlier offer non-binding recommendations and serve as an important instrument in the harmonization of CSTO and CIS states' national legislation. On the one hand, Belarus actively uses all the Model Laws, on the other, it is also known to have provided its domestic legislation as a basis for Model Laws.

According to the CSTO Collective Security Strategy for the period till 2025, one of the CSTO goals in the field of countering international threats and challenges is "to reduce the level of damage to territory and critical objects in the wake of natural and man-made emergency situations" and "to provide conditions for functioning of critical objects in accordance with the national legislature and based on the respective international treaties adopted within CSTO" (PA CSTO, 2016).

Another important milestone with regard to CIP was the adoption by the PA CSTO of the Model *Law "On the safety of critical facilities"* in 2018. In their speeches at the Assembly, the lawmakers stressed the lack of regulation in the legislation of the member states regarding CO and pointed toward the need to improve the quality of the existing normative framework. According to officials' statements, the number of "critical facilities within the economic and social infrastructure system" of the CSTO member countries is constantly growing and the evolving nature of threats presents a continued challenge to member states (Parliamentary Assembly of the Collective Security Treaty Organisation, 2018).

12.2. Implementation

Currently, there is no coordination government body in Belarus designated to regulate CIP-related issues. In a more general sense, the Belarus Council of Ministers coordinates all the state entities involved in emergency response, emergency preparedness, and protection.

Under the *Law "On industrial safety"* (House of Representatives of Belarus, 2016), two bodies primarily responsible for monitoring the implementation of legislation in the field of hazardous production facilities and dangerous objects safety are the Ministry of Emergency Situations and the Ministry of Defense of the Republic of Belarus. The Department for Supervision of Industrial Safety or the Ministry of Emergency Situations carries out the supervision of hydraulic installations, gas and oil infrastructure, pipeline facilities, and other potentially dangerous objects and coordinates activities and information exchange between state bodies, local bodies and other organizations.

Under the same legislation, oil and gas pipeline operators are required to carry out regular risk assessments and risk analysis and submit the results in the form of a document called the declaration of industrial safety to the authorities. The declaration of industrial safety must specify potential accident risks, vulnerability of each facility and potential damage from an accident. An industrial facility's readiness to localize and mitigate such accidents is studied separately.

The State Control Committee, the MIA, the Presidential Security Service, the Ministry of Defense and the State Committee on Border Troops are directly responsible for conducting counter-terrorism efforts. The Ministry of Communications and Informatization is the main government body coordinating the oversight, implementation and regulation of the telecommunications sector. The system of response to nuclear and radiation accidents is integrated into the National Emergency Response System.

In the ICT sector, the *Decree of the President of October 25, 2011, No. 486 "About some measures for safety of critical informatization objects"* (amended in 2013 and 2019) entrusts the Council of Ministers of the Republic of Belarus with the following responsibilities: establishing the criteria for the identification and designation of informatization objects and systems as critical, adopting the protection procedures, and assessing the potential impact on the political, social, economic, information and ecological security in the wake of a threat to a critical informatization object or piece of infrastructure.

In accordance with the same decree, in 2011, the Operational and Analytical Center under the President[1] (OAC) was created to administer and enforce the regulation. The OAC is tasked with regulating the protection of critical informatization objects and data transfer infrastructure, carrying out inspection work, coordinating technical and cryptographic protection activities of state bodies and other organizations, carrying out registration of critical informatization objects and maintaining the national registry of critical informatization objects.

Pursuant to *Presidential Decree No. 486*, the responsibility to design and put in place security arrangement falls on the operator of the critical information object. The owner/operator is responsible for carrying out criticality and security risk assessments and submitting this information to the OAC. The decision to designate an information object or system as critical and the arrangements to include it into the National Registry are then carried out by the OAC. The object is classified as critical from the moment it is added to the national registry.

The same decree provides for the establishment of the National Registry of critical informatization objects and the procedure for their identification and designation. The procedure and the protection measures are further expanded and clarified by means of the *Presidential Decree of December 9, 2019, No. 449 "On the enhancement of state regulation in the field of information security"*. This decree is intended to update the normative framework for the designation of information objects and systems as critical, facilitate the implementation of the National Security Concept provisions and align them with the provisions of the Information Security Doctrine.

The decree obliges the operators of information systems to submit the object security and protection action plan to the OAC. The decree includes guidance on the required format for owners/operators to use when registering the object or a system and submitting the protection plan. In addition, the decree orders the owner and the OAC to conduct periodic inspections (once a year and once in five years, respectively).

12.3. Public-Private Partnership

The infrastructure of national importance in Belarus, most importantly its energy and industrial infrastructure, was built and long existed as an integral and complementary part of a larger Soviet Union. Today these sectors are characterized by a significant share of state-owned enterprises. The state also maintains a high degree of involvement in and control over the country's economy across all spheres. The gas infrastructure in the country is largely under Russian control, while "Gasprom" is the only supplier of gas to Belarus. Gazprom also owns almost all the pipelines in Belarus.

Notably, the concept of public-private partnership has only recently entered the political and legal lexicon in Belarus, and none of the existing normative acts concerned with PPP make any references to CI or CO:

- *The Resolution No. 49 "On Public-Private Partnership Projects"* of July 27, 2016
- *The Resolution No. 532 "On Measures to Implement Law on PPP"* of July 6, 2016
- *The Law No. 345-Z "On Public Private-Partnerships"* of December 30, 2015
- *The Resolution of the Council of Ministers of No. 508 "On Measures to Implement the Law on PPP"* of May 27, 2014
- *The Law No. 53-Z "On Investments"* of July 12, 2013
- *The Law No. 63-Z "On Concessions"* of July 12, 2013

Under the *Resolution of the Council of Ministers No. 508 "On Interministerial Infrastructure Coordinating Board"*, the Infrastructure Interagency Coordinating Council (IICC) was created to coordinate the development of infrastructure facilities within the framework of the PPP (Council of Ministers of the Republic of Belarus, 2014). IICC is comprised of representatives from several ministries, government bodies, public organizations and business structures. The Council participated in the creation of the PPP domestic normative framework and took part in the drafting of the Public-Private Partnership Law.

Article 6 of the *Public-Private Partnership Law No. 345-Z* (House of Representatives, 2015) specifies sectors in which the government may turn to private companies for infrastructure services. The areas include, but are not limited to, transport, defense, gas storage and shipment, electrocommunication, energy and oil refinement, storage and transportation. The Law establishes the grounds for the implementation of a PPP project in these sectors and specifies the process and activities at various stages of the implementation. Notably, energy infrastructure (external supply systems, heat supply systems, gas supply systems) and transport infrastructure (road, rail, air and pipeline transport) are listed first and second on the list of main directions of infrastructure development using the PPP mechanism.

The needs and goals of the PPP project's development are set out in some detail in article 2 of the Resolution No. 508 to prevent the deterioration of infrastructure, attract funds for renovation, improve infrastructure quality, and, crucially, security and infrastructure protection capability against terrorism and other threats (Council of Ministers of the Republic of Belarus, 2014).

National Infrastructure Strategy for 2016–2030, approved in 2015, was initially focused on PPP projects in social infrastructure but was updated in 2017 to include important energy and transport infrastructure. The Strategy lists 100 key projects for possible implementation of the PPP mechanism. Although none of these project is specifically related to CI, the experience gained in their implementation can help the government to apply the same model in the CIP sector in the future. UN experts have already given a positive assessment of the preparedness of the Republic of Belarus to implement PPP projects (Sergi, 2019).

12.4. Cooperation with European Commission

In Belarus, human rights violations and other political concerns have been the main obstacle to dialog with the EU on multiple fronts. As a result, the level of cooperation between Belarus and the European Commission as well as other EU bodies is lower than with the rest of EaP countries. This has implications for cooperation on

CI identification, designation and protection as well as for the provision of technical and legislative drafting assistance from EU organizations.

For instance, Belarus cannot participate in the Euronest Parliamentary Assembly, which is prominent for its support to partner countries in matters related to energy infrastructure and cyber-attacks. Moreover, Belarus remains the only country from the EaP that has not ratified the Partnership and Cooperation Agreement (PCA) with the EU.

However, Belarus has been actively taking part in the multilateral formats of the EaP and different forms of cooperation between Belarus and EU institutions still take place in the matters of cybersecurity, CI-related legal and technical assistance, capacity-building, and financial help.

- From 2007 to 2010, the National Technical Committee of Belarus took part in the INOGATE project "Harmonization of technical standards and practices in the oil and gas sector in Eastern Europe and the Caucasus".[2] The project's goal was to support partner countries in adopting international standards in construction, manufacturing, testing, certification, accreditation, operation and maintenance of gas and oil infrastructure and to create awareness of EU standards, rules, certification and accreditation in the gas and oil sectors.
- In 2013, Belarus' experts contributed to the Good Practices Guide on Non-Nuclear Critical Energy Infrastructure Protection (NNCEIP) from terrorist attacks published by OSCE. The document is a guidebook that "provides a framework that encourages the formulation and implementation of appropriate policies and institutional management of cyber security related to NNCEIP" (OSCE, 2013: 7).
- Between 2011 and 2014, Belarus took part in the Cybercrime@EAP project aimed at strengthening partner countries' capacity to respond to cybercrime and harmonizing legislation in cyber domain. Today, Belarus is benefiting from the participation in another joint European Union/Council of Europe project – CyberEast (Council of Europe, 2020). The program supports EaP countries in action on cybercrime.
- Between 2010 and 2014, 296 participants from Belarus' civil protection and disaster management authorities were among the beneficiaries of the PPRD East 2 Training Programme (European Commission, 2019). The initiative was established with a goal to increase disaster risk management capacities of EaP societies and improve the protection of populations from the negative consequences of natural and man-made disasters.

There is also hope to further strengthen the bilateral relationship through the EU–Belarus Partnership Priorities, which are currently being negotiated.

12.5. Education and Training

A set of state and non-state institutions shares the responsibility for developing CIP competences and training experts in Belarus.

Articles 12–15 of the *Law No. 141-Z of the Republic of Belarus "On the Protection of the Population and Territories from Natural and Technogenic Emergency Situations"* state that organizations need to "hold trainings for workers how to protect themselves, to create and have stand-by warning systems, to ensure conducting rescue operations in accordance with the plans of preventing and eliminating emergencies" (House of Representatives of Belarus, 1998).

In accordance with article 14 of the *Doctrine of Information Security of the Republic of Belarus,*

> Individuals, providing information security, cooperation between government bodies, educational institutions and specialized enterprises in selection, training and employment of such personnel, integration of information security topics into educational programmes at all levels, are trained at the modern educational and technological level, undergo special training, retraining and professional development. The state forms an order for such training.
>
> (Security Council of Belarus, 2019)

Pursuant to article 11 of The *Law No. 363-Z of The Republic of Belarus "On industrial safety of hazardous production facilities"*, the operators or owners of hazardous assets and facilities are required to provide training and certification of workers in the field of industrial safety (House of Representatives of Belarus, 2000). Additionally, article 20 of *Presidential Decree No. 449 "On the enhancement of state regulation in the field of information security"* provides for the education of employees dealing with critically important objects of informatization (President of Belarus, 2019). Furthermore, employees working in the area of information protection are required to have higher education in the field of information protection (or to have another higher or vocational education and undergo retraining/advanced training in the field of information protection).

On the nuclear energy front, Belarus has achieved considerable progress. In 2018, the Emergency Preparedness Review (EPREV) was launched in Belarus under the auspices of International Atomic Energy Agency (IAEA). Among other things, the mission was tasked to review the availability of educational institutions for training specialists to monitor and respond to nuclear emergencies. The final report concluded that the nuclear power personal in Belarus receives regular training in emergency preparedness and that appropriate qualification is maintained over time. The responsibility for such training falls on the Ministry of Education, the Ministry of Energy and the Ministry for Emergency Situations. According to the EPREV report, "The legal basis for this is the resolution of the Cabinet of the Republic of Belarus No. 250 of 28 March 2016, Resolution of the Ministry of Emergency Situations No. 158 of 27 June 2016, Resolution of the Ministry of Education No. 106-a of 24 August 2012, and the Law of the Republic of Belarus No. 183-Z of 27 November 2006" (International Atomic Energy Agency, 2018: 30). Additionally, the EPREV mission itself provided an opportunity to exchange best practices and discuss further harmonization of emergency preparedness among IAEA Member States (International Atomic Energy Agency, 2018: 8).

One of the key highest educational institutions in Belarus providing education in the domain of CIP is the University of Civil Protection of the Ministry for Emergency Situations (UCP). The university offers training for experts in emergencies response as well as in the areas of technospheric security, industrial and radiation safety, civil defense and protection from emergencies of natural and man-made character.[3] The university has served as host to a number of regional and international practical conferences, expert talks and joint training with international organizations. In 2018, UCP had held a meeting of the Belarusian-Polish Sub-Committee on the rescue and protection of the population in an emergency situation. And in 2019, a representative of the university took part in the EU-CHEM-REACT 2 program co-financed by the EU. The program addressed "a need to improve prevention, preparedness and response to chemical disasters in Belarus, Moldova and Ukraine" through a series of best practices exchanges, training, exercises and other activities (The International Centre for Chemical Safety and Security, 2019: 1). The university has a track record of cooperation with the Organisation for Prohibition of Chemical Weapons (OPCW), International Civil Defence Organisation (ICDO), Interstate Aviation Committee (IAC), European Commission (EC) and IAEA.

The International Rescuers Training Centre is another educational facility offering different types of advanced training for the employees of the state emergency response agencies. The center claims to have hosted foreign trainees from more than 69 countries and established cooperation with experts and colleagues from all over the world.[4]

Chapter Conclusion

Following its independence in 1991, Belarus has demonstrated the political will to address the constantly evolving threats to security and willingness to cooperate with the EU on multiple matters. Belarus is relatively prominent for the energy security of EU states and its ability to ensure the safety of oil and gas infrastructure assets is not just a matter of domestic security but is critical for the broader CIS region and for the EU, where some of the final consumers are.

Belarus has been making considerable progress in the area of information systems security and the cyber domain. Belarus has developed a functioning system of coordination on critical informatization objects. Accountability frameworks for critical informatization objects stakeholders have been put in place and National Registry for critical informatization objects is well established.

On the other hand, policy measures that fall into the broader domain of CIP in other sectors remain ill-coordinated and the existing legal base does not adequately address challenges posed to CI safety. These shortcomings put Belarus significantly behind the international curve. Against this backdrop, defining and designating critical infrastructure should be the most immediate step, followed by the elaboration of a unified comprehensive CIP Law in view of best practices of the EU.

Deepening relations between the EU and Belarus and stronger involvement of Belarus in the common European legal space could help the country move in the right direction. CIP agenda crosses national and regional boundaries. Therefore, it is in the interest of the EU to take a more proactive stance, establishing a formal framework for discussing CIP issues with Belarus and to stepping up its legal and technical assistance.

Notes

1. https://oac.gov.by/.
2. INOGATE is a technical assistance and cooperation program between the EU and 11 partnering countries (http://www.inogate.org/).
3. https://ucp.by/en/.
4. https://rescuetraining.ucp.by/eng/index.php/en/.

Chapter 13

Moldova

Denis Cenusa

Contents

This chapter opens up Moldova's legislative and policymaking in the field of critical infrastructure. The country's approach toward critical infrastructure stems primarily from the national documents that conceptualize varying dimensions of security. Apart from strategies that have direct relevance to security issues, there are legal acts that refer to civil protection of the state in extreme situations or focus on environmental contamination, biological security, industrial object, and so on. Whereas these laws reveal the state's motivation to take action against sectoral insecurities, none of them assumes clear-cut functions of protecting the critical infrastructure. The available legal piece dealing with the critical infrastructure stems from the antiterrorism legislation. Consequently, the applied legal approaches are insufficient to prevent the damage of vital assets by hostile internal or external actions.

DOI: 10.4324/9781003273769-16

13.1. Introduction

To understand the real state of play, this subchapter provides a short examination of the existing legislation that addresses the critical infrastructure in a way or another. Depicting the relevant strategic documents or organic laws and normative acts helps understand the efforts and obstacles that the regulation of critical infrastructure is actually facing in Moldova.

Strategies

The first strategic document related to critical infrastructure was the short-lived NSS, which was drafted during 2015–2016, with the involvement of NATO experts. The Supreme Security Council led by the then President Nicolae Timofti approved the NSS in June 2016. But the new president Igor Dodon (2016–2020) with pro-Russian sympathies withdrew the Strategy via the repealing procedure one year later (in June 2017). The Strategy underlined the links with NATO and indirectly viewed Russia as a source of risk. Unlike the old 2011 NSS, the 2016 document reckoned that the critical infrastructure is under external pressure that impairs its modernization.

In the absence of an updated NSS, the document that initiated a firmer discussion about the critical infrastructure constitutes the *National Defense Strategy* (NDS), adopted together with a four-year timeframe 2018–2022 Action Plan, in 2018. The NDS incorporates the notion of critical infrastructure. This is featured as the conglomerate of "devices, networks, services, systems of material objects from the energy, transportation, communication, information technology and utility providing, of strategic interest and public utility". Their destruction or degrading can negatively affect public health and safety, the environment, the functioning of the economy and the activity of state institutions, at the national or local levels. Acknowledging the diverse composition of threats, including the pervasiveness of the hybrid forms of insecurity, the NDS sets out (1) the attack against the critical infrastructure and (2) the cyber-attacks as two of 10 areas of *threats and risks* (Moldovan Parliament, 2018a).

Another document of strategic orientation like the *2017–2019 NATO-Moldova Action Plan (IPAP)*, enacted in 2010, prioritizes the role of defense cyber and points out less exhaustively and profound the critical infrastructure protection (Moldovan Government, 2010).

The more sector-oriented strategic documents mark a keen interest in cyber defense. This includes the *2017 Concept of Information Security*, the *National Cyber Security Program of the Republic of Moldova for 2016–2020* and the *Informational Security Strategy 2019–2024*. All of them direct the attention to the hybrid threats and how these can affect the cyber-related critical infrastructure.

Organic Laws

The legal framing of the protection of critical infrastructure came into being in Moldova through the most recent interpretation of terrorism insecurities. In this sense, the 2017 specialized *Law on preventing and combating terrorism* is the first organic law that terms the critical infrastructure. The origins of this Law are remotely adjacent to European norms. The Law defines the critical infrastructure as

> element, system or component thereof, . . . which is essential for maintaining the vital functions of society, health, safety, security and social and economic well-being of persons and whose disturbance or destruction would significant impact at national level due to the inability to maintain those functions.

This Law establishes that the Anti-Terrorism Center (ATC) of the Intelligence and Security Service (ISS) carries out measures of keeping safe the national strategic infrastructural assets (Moldovan Parliament, 2017). The ATC not only verifies the degree of protection of the infrastructure but also can formulate recommendations to enhance security. This Law overlooks the 2008 Directive on critical infrastructure, though refers to the 2016 Directive on prevention and prosecution of terrorist acts.

In the following part, this chapter will examine the particularities of the present legislation that addresses critical infrastructure. The recent legislative initiatives aimed at instituting a systemic vision about the protection of the critical infrastructure are analyzed critically to establish the trends of the policy-making and political decision-making in the field.

13.2. Implementation

Due to the fragmented nature of the legal framework on critical infrastructure, this subchapter maps out the existing and intended legislation which points to the following two facets: (1) the non-cyber-related infrastructure and (2) the cyber-related infrastructure.

The Non-cyber-Related Infrastructure

The 2016 NDS and the Action Plan codify security perils concerning the critical infrastructure as a whole. The documents exhibit the objects of critical infrastructure that correspond to such spheres of social-economic activity like energy (high voltage power lines), water management (dams), transportation (bridges, viaducts) and communication and digital (hardware networks). The 2018–2022 Action Plan indicates the goal of setting up an approach toward critical infrastructure (Moldovan Parliament, 2018a). As part of the 2018–2022 Action Plan of the NDS,

Table 13.1 The Mentioning of (Geopolitical) Critical Infrastructure in the Major Legal Acts Governing the Critical Infrastructure (Yes/No)

Nr.	Legal Acts	Status	Non-Cyber Infrastructure	Cyber Infrastructure	Geopolitical Connotation
1.	2017–2019 NATO-Moldova IPAP (2010)	Enacted	Yes	Yes	NATO
2.	Draft bill of the National Security Strategy (2016)	Repealed	Yes	Yes	NATO, EU Russia (implied)
3.	National Defense Strategy (2016)	Enacted	Yes	Yes	EU
4.	Law on preventing and combating terrorism (2017)	Enacted	Yes	Yes	External/neutral
5.	Information Security Strategy (2018)	Enacted	Yes	Yes	EU Russia (implied)
6.	Draft law on the protection of essential infrastructure (2020)	Under examination	Yes	Yes	EU Russia (implied)

Source: Author's compilation

the government adopted a series of secondary legislation that put the foundation of a mechanism of protection of the critical infrastructure, under the coordination of the ISS. Thus, the latter has turned into the main institution that coordinates the implementation of the normative acts adopted in 2018 that develop further the 2017 legislation of anti-terror security of the critical infrastructure.

Of vital importance is the *Regulation on the protection of critical infrastructure against terrorism* that entitles the ISS to design and approve the *National nomenclature of critical infrastructure* (Moldovan Government, 2018a). This Regulation enables the public authorities and private entities, also called *operators*, to propose the list of assets to be included in the Nomenclature. The operators are obliged to identify the assets attributable to critical infrastructure. In this sense, a methodology provides three criteria related to the damages deriving from attacks on infrastructure (casualty among the population, size of the economic costs, and the timeframe of the technological disruption). The ATC decides the composition of the Nomenclature and issues an "anti-terrorist passport". While the Nomenclature is not public, the authorities published the sectors that qualify as critical infrastructure counting 11 sectors: *energy, information and communication technology; water supply; food safety; health care; financial sector; national security; administration; transport; chemical, biological and nuclear industry; places of mass gathering of the population*. The whole cabinet of ministers (nine institutions), and other public authorities, manages these sectors. The legislation avoids making reference to the assets of critical infrastructure (power production, gas pipelines) located in the country's separatist region (Transnistria), which is controlled by Russia.[1]

Another important normative act is the *Regulation concerning anti-terrorist testing*. It provides that the intelligence service (ISS) checks the vulnerability of the assets both against the potential object ("weapon") used to produce the damage and the subject ("perpetrator"). The types of tests vary between planned or ad-hoc and opened, dissimulated (confidentiality), and combined (Moldovan Government, 2018b). To ensure the implementation of the recommendations, the ATC may conduct repetitive testing. The outcomes of the testing are published sporadically and in a superficial manner. In the first half of 2020, the ATC carried out tests on CI of water supply (Chisinau capital, Criuleni administrative-territorial unit), transport (postal and luggage transfer), and energy sector. The results of these tests are closed to the public.

An attempt to adopt more specific legislation represents the draft law on *the protection of essential (not critical) infrastructure objects for ensuring national security and public order*. The author of this draft law has espoused the EU *Directive 2008/114/EC* on critical infrastructure and the 2019 *Regulation on screening of foreign direct investments*. None of these EU legal acts is part of the EU-Moldova Association Agreement. Moreover, the government and other institutions involved in the law-making process gave negative opinions upon the draft bill elaborated by the opposition. The Parliamentary Secretariat of the Parliament argued that the draft bill creates duality with the existing laws (public property, lands) and includes various vague provisions. The National Anti-Corruption Centre (NAC) offered the most extensive list of objections concerning the legal wording, lack of financial coverage and provisions that cause eventual corruption risks. The NAC's opinion reveals that the draft bill is oriented to a particular undefined third state that commits "hostile actions" not only against Moldova, but also against the EU, the UK, the US and other Western countries. Currently, Russia matches these characteristics, thereby being targeted by the draft bill. The critique against the draft bill found allies among the Western-related business associations – the American Chamber of Commerce and the EU Business Association. They also warned against the adverse effects on the business environment, investment attraction and property rights. Furthermore, these business associations expressed disapproval of the distorted interpretation of the EU's 2008 Directive and 2019 Regulation.

The Cyber-Related Infrastructure

Important systemic work for cyber protection is carried out under the *2016–2019 National Cyber Security Program*, which seeks to securitize the cyber informational services and improve cyber security management, as a whole. A more strategic approach has the *2019–2024 Information Security Strategy* that prioritizes the establishment of Coordinating Council for ensuring information security and delineates the importance of protecting the informational critical infrastructure (Moldovan Parliament, 2018b). The Ministry of Economy and Infrastructure is measuring the progress of the implementation of the Cyber Security Program, while the Security and Information Service deals with the Information Security Strategy. Thus, between 2017 and 2019, the registered progress included the adoption of the minimum requirements for cyber security in 2017. The new rules apply various tools and filters for building up cyber security through multifactor authentication, firewall, malware protection, antispyware and intrusion test. These rules are mandatory for the central authorities.

Though the government established a reaction center against cyber security incidents (CERT-GOV-MD) by complying with the 2017–2019 IPAP NATO-Moldova, this unit does not function as a national Centre for Cyber Security (CERT), whose creation has been delayed since 2016. The major problems are the incomplete normative setup and financial insufficiency, which makes it impossible to develop a "functional national alert system and real-time information about cyber security incidents". Gradually, the CERT-GOV-MD has undertaken some of the functions of a national CERT. Still, the entities unsubordinated to the government-linked to local public administration and the private sector, as well as their infrastructure – remain uncovered in terms of cyber security. The Ministry of Interior Affairs made attempts to establish its own CERT, but the deficit of finances and the incomplete nationwide CERT led to the postponement. The same deficiencies affect the institution of a similar unit in the military sector.

13.3. Public-Private Partnership

The existing legislation on public property regulates in an extensive manner the relations between public and private sectors. According to the *Law on public property, privatization of the goods of public property* has a determined range of action. There are 14 categories of such assets that are excluded from the circuit of privatization (Moldovan Parliament, 2007). The legislation on the administration of public property encloses a permanently updated list of assets related to national security, national heritage, state reserves, natural resources, transport communication, multi-profile state enterprises, and so on. Under this legislation, the protection from privatization has eight Joint-stock companies and 161 State and municipal enterprises, which are included in an official "List". Furthermore, the lands that are connected to specific economic activity (agriculture) have legal protection from privatization.

The fulfillment of the obligation of the anti-terror legislation is not the only way in which the private and the public sectors operate with the critical infrastructure. To a certain extent, they have to align also with the defense rationality of the state. Thus, the 2002 *Law on the requisition of goods and provision of services in the public interest* obliges the economic entities, along with other subjects (public institutions and individuals), to temporarily cede their (i)mobile properties ("goods") to the state. The situations under which the authorities gain the right to initiate the requisition of the private property constitute the situations emergency, siege and war situations (Moldovan Parliament, 2002).

13.4. Cooperation with European Commission

Overall, Moldova's cooperation with the European Commission does not involve in any consistent way the *Council Directive 2008/114/EC*. Not even the draft bill on "essential infrastructure", blocked in the legislative procedures, is envisioning a specialized interaction with the Commission concerning the critical infrastructure. Nevertheless, the conventional dialogue with the EU rests on the multi-sectoral Association Agreement. This chapter unveils that the Agreement can absorb some deviations due to the Association Agenda's more flexible content that may encompass novel directions of the EU acquis.

Whereas the deepening integration with the EU features a certain sector-by-sector engagement with the critical infrastructure, this engagement seems fragmented because it avoids *Council Directive 2008/114/EC* "on identification and designation of ECI and the assessment of the need to improve their protection". Therefore, by the end of 2020, the EU-Moldova relations did not entail the critical infrastructure in a specific way, unless that was part of the traditional sector-based cooperation as shown further.

The availability of cooperation concerning the critical infrastructure lies in the landmark EU-Moldova Association Agreement and the operational documents guiding and measuring its implementation – the Association Agenda and the Association Implementation reporting (Emerson & Cenusa, 2018). These documents reveal political interest in upgrading various parts of the critical infrastructure, however outside the realm of the *Council Directive 2008/114/EC*. In some cases, the operational documents set out higher ambitions of such cooperation than the Association Agreement itself and the most intense exchange occurs at the level of senior official level. The meetings at this level gather within one of the two subcommittees operational since December 2014 – on Freedom, Security and Justice and Economic cooperation. Otherwise, the EU available engages with the EaP states through programs, such as the EU4Digital or EU4Energy. They complement the EU's bilateral endeavors of supporting the development of Moldova's state policies, in legal or technical terms, in the field of sectoral infrastructures. These programs are contributing to the critical infrastructure, even if in a more limited format.

Through EU4Digital program, the EU backs the cooperation in the field of cybersecurity. Moldova like other EaP countries is eligible for assistance through 2019–2022 "Cybersecurity East" project (EUR 3.1 billion), which combines three types of activities: adopting a legal framework to cultivate proper cybersecurity governance; assisting technically the development of CII; and fostering the operational capacities in situations of cybersecurity disruptions. In this sense, with the EU assistance is willing to finance the training efforts for the national systems of critical information protection, as well as through joint cyber operations and transnational cooperation between the existing networks. In the margin of "Cybersecurity East", Moldova organized the "Cyber Week" online event in November 2020, with officials and professionals, including from the EaP countries. The event included webinars on cyber incident response and cyber risk management. The cooperation concerning the countering of cybercrime of more systematic character envisages the strategic and operational agreement between Moldova and one of the EU's law enforcement agencies – Europol. To this end, the national contact point in Moldova ensures 24 hours exchange of information with Europol on investigating a range of offenses, including the "computer crimes".[2]

In the non-cyber area, the cooperation with the EU institutions take place via programs like EU4Energy that supports financially the Energy Community's Secretariat to improve the energy governance in Moldova. That serves for enabling secondary legislation with which the national regulatory body promotes better functioning energy-related critical infrastructure. Similarly, cooperation evolves in the border management domain. The EU Mission (EUBAM) at the Ukraine-Moldova common border has an instrumental role not only eradicating the irregularities. It has extended six times since 2005. The EUBAM actively participates in implementing the objective of integrated border management[3]. Consequently, the old cross border checking points undergo renovations and new joint Ukraine-Moldova crossing points, such as Palanka-Mayaki-Udobnoe crossing point (from 2019), come into being.

13.5. Education and Training

There is no systemized format of education or training in the field of critical infrastructure that involve the public servants and institution. The 2017 anti-terror ISS shall organize training for the operators (both public and private sectors) of the assets that are included in the Nomenclature of critical infrastructure. In terms of cyber-related infrastructure, the Public Administration Academy that prepares public servants to serve the government's priorities conducts courses that are annual state's requests ("state orders"[4]). Only the courses on e-governance offered to develop digital diligence remotely resemble the cyber issues. Therefore, the diffuse of knowledge on cyber-security, existent in specialized departments and state agencies, to public servants in an extensive manner is lacking and overlooked. Even less so is done in the field of critical infrastructure as a whole, which requires specific sector-related backgrounds (transportation, energy, etc.). As far as it concerns the educational system, the Ministry of Education, Culture and Research is in charge of drafting the school curriculum in order to introduce cyber security disciplines. Progress in the area was not reported as for January 2020.

Furthermore, in the field of cyber and informational field, the Moldovan officials take part in various activities of expertise exchange and learning: with the support of the US on cyber security – "CyberSec 2017"; at the initiative of the International Telecommunication Union "2017 Common cyber drill ITU – ALERT for Europe and CIS regions"; on cyber criminality together with Council of Europe's Cybercrime Program Office – Cybecrime@EAP 2018. Furthermore, the representatives of the government's CERT (ITSec) also participated in specific regional conferences and seminars dedicated to the management of information security ("7th Regional Cybersecurity Summit") or more generally to cyber security (2018 Geneva Information Security Day, NatCSIRT 2018 by the Carnegie Mellon University). Within the Science for Peace and Security NATO program, the Research and Training Laboratory in the field of cyber security has been launched in 2016 and another Cyber Security Laboratory in 2018 at the Technical University in Moldova.

Since 2019, the Moldovan research institutions initiated collaboration dialogue within the European Cooperation in Science and Technology (COST). The training and education component, primarily in the cyber area, that provides EU4Digital program is concentrated in "Cybersecurity East" project. The latter has fostered communication and exchange of knowledge between the EU and the

EaP countries, including Moldova. Thus, the Moldovan government's employees (Ministry of Economy and Infrastructure) benefited from the training organized during the "European Cyber Security Month" in October 2019. In 2020, Moldovan law enforcement bodies, the national CSIRTs and academia joined the training co-organized by the EU and the Council of Europe focusing on critical information systems linked to electoral processes. Overall, the knowledge-building activities held within "Cybersecurity East" are enabling the nurturing of cyber skills. However, they are not either systemic or comprehensive and depend on the self-organization of the national government.

Chapter Conclusion

Whereas the Moldovan authorities commonly express determination and consistent concentration on cyber defense, other sides of critical infrastructure are addressed in a dispersed and incomplete manner. This derives from a lack of holistic vision about the interconnectedness and interdependence of the national infrastructure. The systematization of the approaches, which have developed in the last decade, appears to be the essential requirement to build sound vision, primary and secondary legislation to establish a politically neutral and durable mechanism for protecting the national critical infrastructure.

The most proximate and adequate modus operandi is available in the EU, where the Member States are applying the *Council Directive 2008/114/EC* on critical infrastructure. The first step in this direction has to be a skillful transposition of the EU legislation that would exclude any legal controversy and suspicions of corruption risks or of damages oriented against the business interests.

Notes

1. Expert-Forum, "War by other means. Kremlin's Energy Policy as a Channel of Influence. A comparative assessment. Case studies from Ukraine, Moldova, Georgia, Romania and Hungary", 2019.
2. Europol-Moldova Agreement on Operational and Strategic Cooperation, April 2014.
3. EU, Association Implementation Report on Moldova, September 2019.
4. Moldova's Government approved one time per semester the Plan for Professional Developments of the Public Officers at the State's Request in line with the Law nr. 158 (Art. 37) of 2008 on public position the status of the public servants.

Chapter 14

Georgia

Grigol Modebadze

Contents

This chapter reviews the historical experience of Georgia and the existing legislation on critical infrastructure, as well as provides information on newly started reform within the field of CIP and shares the main content of the proposed law of Georgia on CIP.

14.1. Introduction

For the timely and adequate reaction to sustainable socio-economic development and threats of the country, it is especially important to define and protect the

DOI: 10.4324/9781003273769-17

objects, services, systems and their parts that are functional and strategically influential for the state. If those mentioned earlier are being destroyed or demolished then the threat toward national interests will significantly increase, national security will also be affected, lives of citizens and their well-being, as well as the socio-economic state, will all fall under a significant threat.

Georgia has no formally defined list of critical infrastructures; it is not regulated by legislation and there is no identification, division of responsibilities and protection of critical infrastructure. Even though there is no term called critical infrastructure (excluding CII), which is regulated by the *Law of Georgia on "Information Security"* (Parliament of Georgia, 2012) in Georgian legislation, we can meet several related terms, such as objects of strategic and special importance and property of the special importance. The Georgian legislation, within the field of CIP and defining the list of objects of special importance, has gone through many changes. So, it is very important to overview the changes in the Georgian legal framework and introduce those laws which refer to the areas of CIP.

Law of Georgian on "State-Owned Property" defines the Government of Georgia as the body that approves the list of objects of special importance (Parliament of Georgia, 2010). The concept of objects of strategic and special importance was defined by the *Law of Georgia on "Engineering-Geodesic Control and Objects of Strategic and Special Importance"* (Parliament of Georgia, 2003). Buildings, which are functionally and strategically influential for the defense and security of the country, territorial-spatial arrangement, cultural heritage, economics, environment, and natural resources are classified as the objects of strategic and special importance. The list of the special objects was approved by the Government of Georgia on the basis of the *Law on "Privatization of State Property, Privatization of Property of Local Self-Government Unit and Transfer of Right to Use"* (Parliament of Georgia, 1997).

The list of objects of strategic and special importance was defined in 2003 by the Minister of Urbanization and Construction with the *Decree №58 on "the List of objects of strategic and special importance"*. The list consisted of 21 categories: main oil pipeline, pipeline, reservoir, high-voltage power line, hydroelectric, thermal and nuclear power plants, main roadway, building of central TV, all TV and communication towers, bridges, tunnels, metropolitan, railway station, airport and marine port buildings, concert halls, libraries, high educational institutions, and university buildings, sports and medical institutions.

The owner of the objects of strategic and special importance was obliged to organize the safety measures of the object; at the same time, the protection and supervision of objects was carried out within their competence by the Ministries of Defense, Internal Affairs, State Security Services, State Departments of State Border Protection and Intelligence, and the Special State Protection Service. In addition, the protection of confidentiality and state secrets was regulated by the *Law of Georgia on "State Secrets"* (Parliament of Georgia, 1996a).

In 2008, by the *Decree №93 on the "Property of the special importance"*, the list of strategic facilities decreased. The term has been changed and only the "property of the special importance" remained as a title. Based on the decree, the objects that were put on the list (Government of Georgia, 2008) were airport takeoff and landing lines (Senaki and Marneuli) and berth of harbors, waterworks, traffic lights, and water area (Poti and Batumi). In 2010, the aforementioned Law of 2008 Decree №93 was declared invalid and the list of Properties of Special Importance was confirmed and regulated by the *Law on "State Property"*. Article 4, paragraph 6 of the Law states that the Government of Georgia asserts the list of state property of the special importance.

In 2014, at the Economic Council meeting, with the initiative of the prime minister, the government was given the assignment to start working on the *Law of Georgia on "Property of Special Importance"*, namely, on updating the list of objects of strategic importance (as the concept of objects of strategic importance no longer existed and the list was abolished) and presenting it to the Parliament for approval[1]. However, no changes have been initiated.

In the last decade, the state has established a set of legal documents which refer to the areas of CIP, such as:

- *Law of Georgia on "State Owned Property"* (State Property which is not subject to privatization) (Parliament of Georgia, 2010)
- *Decree №584 of the Government of Georgia on "the List of Subjects with High Risk for State Security"* (2015)
- *Decree №312 of the Government of Georgia on the approval of "the List of Critical Information System Subjects"* (2014)
- *Law of Georgia on "Information Security"* (2012)
- *Decree №567 of the Government of Georgia on approval of "List of Critical Information System Subjects in Defense Field"* (2014)
- *Order №427 of the Government of Georgia on "the Approval of the List of Persons and Items of Protection of Special State Protection Service"* (2013)
- *Order №1–1/595 of the Minister of Economic Development of Georgia, on "the List of the Property of Special Importance in the Civil Aviation Sector of Georgia"* (2010)
- *Law of Georgia on "Promotion and Guarantees of Investment Activity"* (1996)
- *State Property which Is Not Subject to Privatization (Law of Georgia on "State Owned Property"* [Parliament of Georgia, 2010])

The state property that is not subject to privatization includes earth, minerals, water resources, territorial waters, continental shelf, state forest funds, airspace and means of control and management of air traffic, state reserves, natural monuments, objects of historical, cultural and artistic values, religious and cultic buildings, takeoff and landing lines, state archives of historical and cultural significance, state funds for cinema and phonographs, museum collection and house-museums, berth of harbors, roads, frequency spectrum, Georgia's position on geostationary orbit, state pantheons, and state-owned special agricultural lands (Parliament of Georgia, 2010).

- *List of Subjects with High Risk for State Security (Decree №584 of 2015 of the Government of Georgia on Approval of "the List of Subjects with High Risk for State Security")*

The list of subjects with a high risk for state security has been defined based on article 22, paragraph 4 of article 51 and paragraph "b.b" of *Law on "State Security Service of Georgia"*. The list consists of 23 subjects: ministries and legal entities of public law which are under their sub-entities (e.g. LEPL – State Services Development Agency of the Ministry of Justice of Georgia), city halls of major municipalities (e.g. Batumi), large state-owned companies (e.g. JSC "Georgian Railway" Ltd, "Engurhesi"), airports (e.g. "Tbilisi Shota Rustaveli International Airport"), ports (e.g. Poti Port), and private and state-owned companies (e.g. JSC "Telasi").

The State Security Service of Georgia (SSSG) is responsible for ensuring security in entities with high risk for security of the state. Consequently, it is authorized to establish safety regimes and control their protection. Also, SSSG is authorized to define necessary measures to ensure security and the rules and conditions of monitoring the fulfillment of the requirements set by the security regime (Government of Georgia, 2015).

- *List of Critical Information System Subjects (Decree №312 of 29 April 2014 of the Government of Georgia on the Approval of "the List of Critical Information System Subjects")*

The list of critical information system subjects was defined based on article 3 paragraph 2 and article 11 paragraph 3 of the Georgian *Law on "Information Security"*. The list was developed by the Ministry of Justice in cooperation with the MIA, the Ministry of Defence and the SSSG. The list consists of 39 subjects, which included ministries and their sub-entities – legal entities of public law (e.g. Ministry of Finance of Georgia and LEPL – Revenue Service), Tbilisi City Hall, large state-owned companies (e.g. JSC "Georgian Railway", Ltd "Union of Georgia Airports"), National Bank of Georgia, The Parliament of Georgia, Administration of the President of Georgia and others (Government of Georgia, 2014a).

The LEPL operating under the Ministry of Justice of Georgia – Data Exchange Agency – is responsible for information security. Therefore, in relation to the critical information system subjects, it establishes the minimum information security requirements (considering the critical nature of critical information system subjects) in accordance with the standards and requirements set by the International

Organization for Standardization (ISO) and Information Systems Audit and Control Association (ISACA). Namely, the DEA establishes the rules of management of information assets and carries out the compliance of the minimum level of security (Information Security Audit). Incidental management and other activities related to the information security coordination are performed by the DEA – Computer Incident Response Group – CERT.GOV.GE.

- *Law of Georgia on "Information Security"*

This Law aims to promote the efficient and effective maintenance of information security, define rights and responsibilities for public and private sectors in the field of information security maintenance, and identify the mechanisms for exercising state control over the implementation of information security policy (Parliament of Georgia, 2012).

- *List of Critical Information System Subjects in Defense Field (Decree №567 of 29 September 2014 of the Government of Georgia on Approval of "List of Critical Information System Subjects in Defense Field")*

The list of critical information system subjects in the defense field and the project on the classification of the critical nature of relevant subjects will be developed by the Ministry of Justice in cooperation with the MIA, the Ministry of Defence and the SSSG. The list of critical information system subjects in the field of defense was defined on the basis of article 101 paragraph 4 of the *Law of Georgia on "Information Security"*. The list consists of five subjects: The Ministry of Defence and four sub-entities of the ministry (e.g. LEPL "Cyber Security Bureau") (Government of Georgia, 2014b).

Information security requirements for the critical information system subjects in the defense field are determined by Cyber Security Bureau in accordance with the standards and requirements of the ISO and the ISACA.

Cyber-attack on critical information system subjects in the defense field, incident handling and other preventive activities are carried out by the DEA – Computer Incident Assistance Group – CERT.GOV.GE.

- *The List of the Items Protected by the Special State Protection Service (Order №427 of May 17, 2013, of the Government of Georgia on "The Approval of the List of Persons and Items of Protection of Special State Protection Service")*

The list of the items protected by the Special State Protection Service was defined by article 5 paragraph 7 of the *Law of Georgia on "Special State Protection Service"*. The list consists of 21 subjects: administrative buildings of the Government of the Autonomous Republic of Adjara and Supreme Council of Adjara, residential apartments of the chairpersons of the Supreme and Constitutional Court of Georgia, prime minister and president's residence apartments, and Georgian Patriarchate (Government of Georgia, 2013).

The Special State Protection Service, in agreement with the administration of the objects of protection, defines the system of access to these facilities and the protection measures, gives recommendations to the administration of the objects of protection by equipping them with technical means of protection of the buildings (protection object) in the municipality.

- *The List of the Property of Special Importance in the Civil Aviation Sector of Georgia (Order №1–1 / 595 of the Minister of Economic Development of Georgia, April 23, 2010, on "The LIST of the Property of Special Importance in the Civil Aviation Sector of Georgia")*

The list of the property of special importance in the civil aviation sector of Georgia was defined by article 7 of the Georgian Air Code and article 13 paragraph 1 of the *Law of Georgia on "Normative Acts"*. The list consists of eight subjects: airports of big cities (Tbilisi, Batumi, Kutaisi and Sokhumi), Georgian airspace, international air routes and local lines, air traffic service/management techniques, and helipads (Mestia and Lagodekhi) (Minister of Economic Development of Georgia, 2010).

Ensuring the safe operation and aviation security of civil aviation, as well as the promotion of international standards in the field of civil aviation, is carried out by the Civil Aviation Agency, which is the LEPL under the Ministry of Economy and Sustainable Development of Georgia.

- *Law of Georgia on "Promotion and Guarantees of Investment Activity" (1996)*

The Law defines internal and external investment policy, as well as guarantees certain security aspects, and also establishes restrictions in terms of investments. The investment restrictions cover the areas such as nuclear, bacteriological, chemical weapon production and distribution, and production of narcotics. The Georgian government is the only body that is allowed to invest in energy dispatching production. In the field of transport and military production, it is forbidden to hold ownership by foreign and private investors (Parliament of Georgia, 1996b).

14.2. Implementation

Ensuring secure and resilient operation of critical infrastructure demands a unified approach, clarity of the goal, mutual sharing of roles and responsibilities by public and private sectors, rational distribution of limited resources, accountability, and effective coordination.

As we mentioned, the current legislation on national security does not cover any legal bases for critical infrastructure security and the existing Georgian legislation does not provide unified regulations on the status of critical infrastructure. In contradiction, as we mentioned earlier, there is a certain "sectoral approach", where subjects are defined by sectors. In addition, there are provisions in the Georgian legislation that define subjects through more "general principles", and their critical status depends on the risk level to the national security (Government of Georgia, 2015).

The most developed sector within the CIP area is the information security sphere in Georgia, where separate regulations are in place (Government of Georgia, 2014a). Due to the fact that information security represents a rather narrow part of critical infrastructure, from a wider perspective high-level information security is not enough for ensuring the security of critical infrastructure on a national scale. It is worth noting that according to Georgian legislation, there should be a defined list of state properties that are of critical importance. *The Law on "State Property"* defines the Georgian government as an authorized body for approval of the list; however, on a state level, there is no such list. As a result, the latter creates a legal vacuum. Therefore, the CIP system is going to be reformed in Georgia.

The aim of the reform described in this document is to create a regulatory framework related to critical infrastructure and the formulation of measures to be taken for its effective implementation. This reform is complex and unites a number of issues. However, the three main objectives can be pointed out:

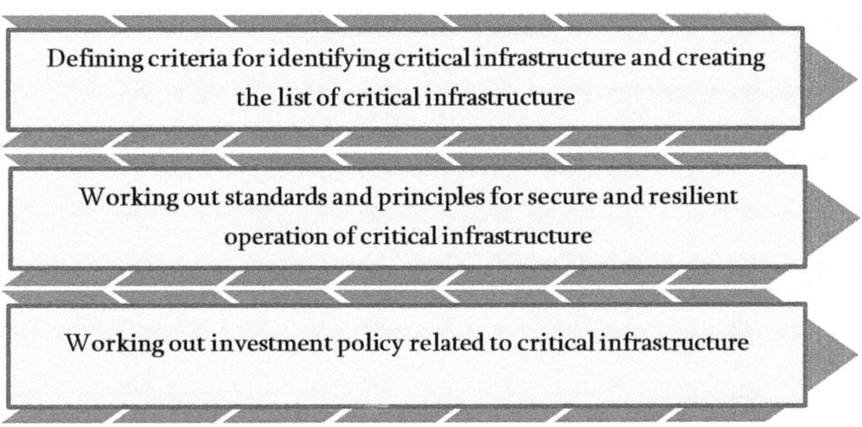

Figure 14.1 Main objectives for CIP framework development

The reform and planned activities are based on experiences of the EU Member States (Slovenia, Czech Republic, Romania and Estonia), UK and USA regarding critical infrastructure as well as principles and recommendations worked out by international organizations, such as NATO, EU and the United Nations.

To come up with an effective model of the regulatory framework for critical infrastructure, this document has been composed considering international best practices (UK and Slovenia), problematic aspects and expert recommendations. This also ensures the harmonization of national legislation with EU law.

It is the primary duty of every government to ensure the resilient operation of essential services and services needed for the full-fledged functioning of institutions. Secure and resilient operation of critical infrastructure has high priority in EU and NATO countries.

The creation of a modern regulatory framework for critical infrastructure in Georgia is directly linked to the harmonization of the country's legal system with European law. It also implies fulfillment of international obligations regarding state security as well as the standards and recommendations that have been worked out by international organizations (NATO, EU and UN) throughout the years.

14.2.1. Defining Criteria for Identifying Critical Infrastructure and Creating the List of Critical Infrastructure

Taking the aforementioned into consideration, the CIP system is going to be implemented through the legal framework of the law, which the Georgian Government is going to adopt soon. Considering international experience and the Georgian reality, certain institutional reorganization is needed for the adoption of the proposed *Law on "Critical Infrastructure Protection"*.

The aim of the proposed *Law on "Critical Infrastructure Protection"* is to identify, determine and protect critical infrastructure as well as define competencies, responsibilities and other issues of importance for the national CIP system. To ensure security and continued functioning of critical infrastructure, it is of utmost importance to define the institutional scheme of critical infrastructure management security. In the case of Georgia, a three-level approach is going to be considered:

- The primary agency with the highest (political) decision-making power will be the Government of Georgia.
- The secondary agency will be the Critical Infrastructure Coordination Agency (Coordination Agency). The coordination agency will work out principles and methodologies of regulation, coordinate and monitor the implementation process.
- The tertiary agency will be a sector-specific agency, a ministry or ministries, which will be in direct communication with the sector-specific critical infrastructure and support implementation of requirements envisaged in the Law.

Proposed institutional organigram

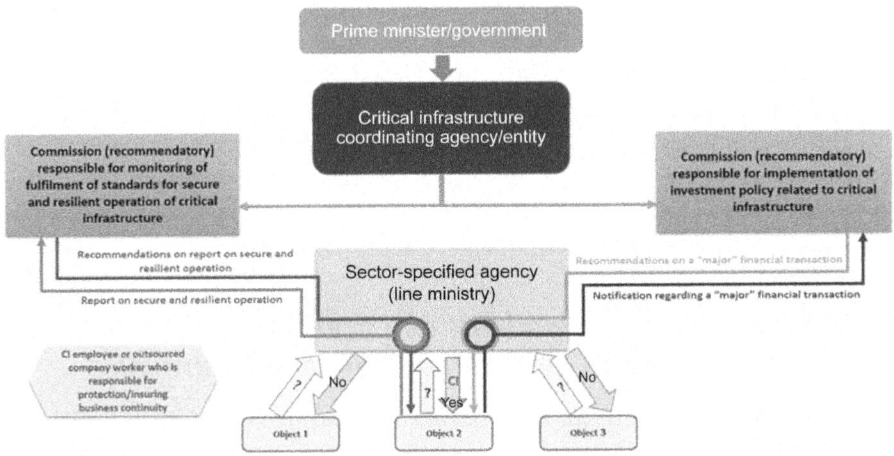

Figure 14.2 CIP proposed institutional organigram

The government has an important organizational role in the coordination of the regulatory framework regarding critical infrastructure. This includes: defining critical infrastructure, coordinating tasks of sector specific agencies, approving critical infrastructure identification criteria's methodology and procedures, approving standards for a secure and resilient operation of critical infrastructure, and approving investment policies related to critical infrastructure and other specific obligations described in the proposed Law on CIP.

According to the proposed Law on CIP, the rights and responsibilities of the institution that coordinates the activities in the field of CIP are: defining the methodology for identifying critical infrastructure; adopting the decision to determine status and category of the critical infrastructure as well as the list of critical infrastructure (based on the recommendation made by the sectoral authority); determining the methodology for assessing the risk of critical infrastructure functionality (with involvement of sectoral agencies in development of the methodology); developing of the standards for safe and continuous functioning of CI with the engagement of sectoral agencies; determining critical infrastructure investment policies; and providing information to the sectoral authorities on threats.

The rights and responsibilities of the Critical Infrastructure Protection Authority in the field of CIP are: participation in the development of a critical infrastructure risk assessment methodology and identifying sector-specific risks; drafting additional measures to ensure the safe and continuous operation of critical infrastructure based on decisions taken/proposed by the government; provision of technical assistance to the critical infrastructure owner/operator to establish standards for the safe and continuous operation of the critical infrastructure; and within its competence, make recommendations to the owner/operator of the critical infrastructure on the account of their safe and continuous functioning. The Sectoral Critical Infrastructure Responsible Authority also designates a contact person(s) who will collaborate with the critical infrastructure owner/operator and other sector critical infrastructure officers in the field of CIP.

The owner/operator of the critical infrastructure shall ensure the safe and continuous operation of the critical infrastructure. They shall designate the contact person(s) who will collaborate with the other critical infrastructure owner/operator and the sector critical infrastructure officer in the field of CIP.

14.2.2. Working Out Standards and Principles for Secure and Resilient Operation of Critical Infrastructure

According to the proposed *Law on "Critical Infrastructure Protection"*, the principles of critical infrastructure protection are as follows:

- *Integrated approach* – It requires the involvement of all responsible agencies and organizations in CIP measures. This includes both preventing critical infrastructure failures and responding to such interruptions. Within the same approach, different types of threats based on the relevant risk assessment should be taken into account, which also implies interdependence of the critical infrastructure sectors and their interconnections.
- *Responsibility* – It involves the obligation of the owner/operator of the critical infrastructure, as well as of all relevant agencies and organizations, to ensure the operation and protection of the critical infrastructure.
- *Protection against various types of threats* – It requires all responsible agencies and organizations to take into account natural and man-made threats to ensure the continuous functioning of critical infrastructure.
- *Continuity of Critical Infrastructure Protection Action Plan* – It requires support for Critical Infrastructure Protection Action Plan by ensuring continuity of critical infrastructure risk assessment process and compliance with safeguards.
- *Information and data sharing* – It requires regular, timely and trust-based exchange of critical infrastructure information and data by all responsible agencies and organizations in accordance with the requirements of the legislation on the protection of classified information.

Regarding the CIP measures:

- In the event of an increased threat, force majeure or crisis, when major measures to protect critical infrastructure are insufficient, additional protection measures may be implemented.

- CIP measures shall be implemented by the owner/operator of the critical infrastructure, taking into account the risk assessment.
- Within its competence, the agency responsible for the operation of sector-critical infrastructure may implement additional safeguards or submit proposals to the government on the feasibility of implementing additional measures to protect critical infrastructure at the sector level.

14.2.3. Working Out Investment Policy Related to Critical Infrastructure

The principles of freedom of investment have been recognized by all developed countries though there are certain investment policies in connection to critical infrastructure. This in itself involves a number of restrictions in relation to the ownership and disposal of assets and service of critical infrastructure.

Georgia is a small-economy country and significantly relies on foreign investments. Therefore, the country's economic policy is primarily focused on attracting and supporting foreign investment in Georgia. Though from a national security perspective it is important to secure the country from investments that carry national-level threats (within specific sectors/specific objects).

There are certain regulations within the Georgian legislation on restrictions and bans of investment in a number of spheres. The *Law of Georgia on "Promotion and Guarantees of Investment Activity"* represents a general framework (Parliament of Georgia, 1996b) as it regulates certain "spheres" and does not control transactions of specific subjects, which might be to some extent related to national security risks. Therefore, there is a need to update the *Law on "Promotion and Guarantees of Investment Activity"* in order to specify certain aspects.

The topic of investment screening is a separate component of the reform within the field of CIP. When general CIP legislation will be adopted, reference will be made to a special screening procedure approved by the government, applicable to an individual investment transaction (qualified financial transaction) with regard to a particular critical infrastructure subject.

Within the institutional context, it is aimed to create a governmental commission (investment monitoring commission), entrusted with the responsibility of studying the respective individual investment transactions, delivering the conclusions/findings and submitting the respective recommendations to the Government of Georgia. It is envisaged that those recommendations will lead to the approval or blocking of the requested transactions. Both foreign direct as well as portfolio investments will go under some scrutiny. The Government of Georgia aims to introduce a legal regime, which, on the one hand, provides for the guarantees (including legal safeguards) for the protection of investment activities, and, on the other hand, ensures the protection of national security interests.

It has to be stressed that with regard to the specific component of investment screening regulation, the Office of the National Security Council is in active communication with the US Embassy Tbilisi, and the respective component will actively be dealt with through the respective US–Georgia bilateral sectoral cooperation channels.

14.3. Public-Private Partnership

The well-being of every state depends on continuous and well-functioning critical infrastructure objects, systems and services. Potential analyses of critical infrastructure reveal that most of the objects fall under private sector ownership. Thus, the responsibility of ensuring the secure and continuous functioning of critical infrastructure falls on the private sector, though the state will also take part in the process.

The Georgian normative model of public-private partnership is regulated by the *Law of Georgia on "Public-Private Partnership"* (Parliament of Georgia, 2018). This Law determines the legal basis for public-private partnerships, including the rules and procedures related to the development and implementation of public-private partnership projects, the principles of public-private partnership, and relevant institutional systems, as well as other matters related to public-private partnerships. Unfortunately, this law does not deal with public-private partnership within the field of critical infrastructure, and its objectives are related to enhancing the efficiency of projects, meeting public interests by creating new public infrastructure and/or providing public services and/or improving existing public infrastructure and public services, attracting private financing, enhancing the efficiency of public finances, arranging the sharing of risks between the public and private sectors, and benefitting from the know-how of a private partner.

As was mentioned earlier, most of the potential critical infrastructures are under private ownership. Thus, the private sector knows all weaknesses and strengths of their subjects and they should take a crucial part in strengthening critical infrastructure resilience and protection. It is worth noting that due to the low level of awareness organizations/regulators find it difficult to communicate with relevant sectors. Meanwhile, some of the organizations and regulators believe that ISO standards that are in place are sufficient and there is no need for establishing further national-level standards, as the latter might have a financial impact on the owners of the objects. Thus, there is not enough awareness that the national-level standards are of higher importance and carry significant security implications in comparison to international standards.

Without a doubt, no nation can achieve resilient functioning of critical infrastructure if there is no tight cooperation between the public and private sectors, especially when we deal with the exchange of sensitive information and data, as well as the development of threat, risk and impact analysis, and contingency plans.

14.4. Cooperation with European Commission

It is the primary duty of every government to ensure the resilient operation of essential services and services needed for the full-fledged functioning of institutions. The secure and resilient operation of critical infrastructure has a high priority in EU and NATO countries.

The creation of a modern regulatory framework for critical infrastructure in Georgia is directly linked to the harmonization of the country's legal system with EU *Acquis Communautaire*. It also implies the fulfillment of international obligations regarding state security as well as of standards and recommendations that have been worked out by international organizations (NATO, EU and UN) throughout the years.

It is relevant to note that according to the EU Association Agreement, Georgia has to harmonize with directive of the EU Commission (2008/114/EC) – on ECI where there is a definition given on critical infrastructure, general identification criteria, demand on establishing secure operator plan, demand on identifying and establishing a list of responsible personnel and points of contact on a state level, and reporting and securing classified information. Georgia has strong aspirations towards EU integration, so it is important to get engaged in the above-aforementioned topics from the early stages.

Significant effort is made by international partners in relation to providing support toward the critical infrastructure reform. And due to increased interest from a wide range of donors, certain communication issues have been identified. Even though NATO–Georgia Professional Development Programme (PDP) and the EU delegation provide support in those areas, individual NATO and EU member states also try to provide additional bilateral support to specific sectors on critical infrastructure.

There are ongoing projects strengthening the resilience of CI. We can mention some of them: in the energy sector, SCADA system and information security management improvement projects are currently ongoing with the support of international partners (US Energy Association and USAID). In addition to the development of cybersecurity within energy production and enhancement of cybersecurity capabilities. Also, in the agriculture sector, as a result of the EU Association agreement, there are continuous strict security regulations in place on phytosanitary inspection.

The National Security Council (NSC) Office, as an agency coordinating the reform on CIP, plans to organize and coordinate elaboration/approval of the comprehensive gap analysis document in the field.

The aim is to acquire baseline data for further implementation of one of the reform working directions. The latter encompasses conceptual as well as legal elements and is essentially aimed at creating a conceptual-legislative framework of the aforementioned reform. Thus, it is planned to elaborate a National Strategy on Critical Infrastructure Protection (CIP) and the general legislative framework in the field. Accordingly, on the basis of the present request, it is aimed to guarantee peer review/assessment of the gap analysis document, updated (elaborated) with the coordination of the NSC Office, in active cooperation with the respective sectoral agencies. At the same time, expert assistance is requested with the aim of delivering the review/assessment of the general draft legislation in the field of CIP. With the aforementioned aspects in mind, and because of the sensitivity and complexity of the topic in question, the Government of Georgia activated the technical assistance and information sharing instrument of the European Commission TAIEX (event #31590).

Also, the Government of Georgia asked for donor assistance through the EU Program/Twinning instrument, which aims to elaborate and implement threat, impact and risk assessment methodologies in the respective sectors, work out the sectoral and national criteria for the identification of critical infrastructure, and determine criticality levels as well as define general or sector-specific measures for the protection of critical infrastructure. To ensure successful elaboration and implementation of threat, impact and risk assessment methodologies in the respective sectors, all agencies have created interagency working groups, and representatives of these groups will participate in any activities organized under the Twinning instrument.

14.5. Education and Training

In order to successfully implement the critical infrastructure reform, it is important to have an understanding and awareness of the topic within each specific sector (organization, regulators and private sector representatives). The topic is relatively new for Georgia, therefore there is limited knowledge within the organizations on critical infrastructure management security of their respective sectors.

In terms of awareness, there are certain exemptions such as National Bank, LEPL, DEA and Georgian National Communication Commission. Those agencies are far advanced in this sphere as they have been taking certain steps on the identification of critical infrastructure and ensuring their security (National Bank is the most advanced as they have already developed threat and risk assessment analysis).

Human resources in each system represent one of the most important components of successful operation. Unfortunately, Georgia does not have a specific program in any higher education institution within the field of CIP.

Chapter Conclusion

Unfortunately, in Georgia, there is no unified legislative framework on critical infrastructure. Though there is a "sectoral approach" in place, there are certain overlaps, as some of the objects of critical infrastructure are under the protection of a number of organizations according to their sectors/functional competencies. According to the existing legislation, there are no unified security standards on critical infrastructure as well.

It is worth noting that, there is a list where subjects are compiled according to rather "general principles" and their critical status is based on their high national security risk factors. Among other relevant subjects, the most relevant aspect is strategic importance, which in itself restricts privatization and investment or the outlined subjects are under the responsibility of the Special State Protection System. The Georgian legislation outlines certain regulations in some spheres and systems in regard to restrictions on investment. There is a general framework that is outlined by the Law of Georgia on *Promotion and Guarantees of Investment Activity* but it requires additional specifications in relation to investment in critical infrastructure.

The most developed sphere in Georgia is information security, and there are separate regulations in place. Information security represents a rather narrow aspect of critical infrastructure, therefore, only advancing information security is not sufficient in terms of CIP. Therefore, in the sphere of critical infrastructure, the advancement of information security only is not sufficient. On the other hand, information security still represents an important and one of the crucial parts of critical infrastructure, specifically while looking at the establishment of the legislative framework. In the framework of critical infrastructure reform, it is worth noting that information security is far advanced and highly developed thus the need to align the reform process to the existing high standards.

Note

1. As mentioned, the Law on State Property defines the list of state strategic objects that is developed by the government.

Chapter 15

Armenia

Ashkhen Beglaryan

Contents

Critical infrastructure protection, including protection of cyber risks, has a strategic role for the Republic of Armenia (RA), as they ensure the availability of vital goods and services. Disruption or destruction of CI can have a serious negative impact on the well-working of the state, private and public sectors. Therefore, in the developing world, and the new global dynamic international experience and cooperation, including the European Union, are always actual for Armenia.

15.1. Introduction

Relations between the European Union and Armenia are based on the *EU-Armenia Comprehensive and Enhanced Partnership Agreement* (CEPA), a modern, ambitious Agreement, which was signed on November 24, 2017, in the margins of the Eastern Partnership Summit in Brussels and entered into provisional application on June 1, 2018, (EU Neighbours Portal, 2020). This Agreement provides a framework for

DOI: 10.4324/9781003273769-18

Armenia and the EU to work together for the benefit of the citizens of Armenia in the areas of strengthening democracy and human rights, creating more jobs and business opportunities, fairer rules, more safety and security, cleaner environment as well as better education and opportunities for research.

The CEPA was signed on behalf of the European Union, of the CEPA between the European Union and the European Atomic Energy Community and their Member States, of the one part, and the RA, of the other part (European Union, European Atomic Energy Community, Republic of Armenia, 2017).

In the academic literature, critical infrastructures (CIs) are considered in their expanded meaning. CIs mean material (for example, electricity or irrigation systems) and intangible (for example, public peace between the government and the people, under a stable "public contract") infrastructure, on which even small-scale external influences or their internal dynamics (fluctuating or deterministic) can lead to large-scale negative repercussions, even to the point of loss of stability and controllability of the given infrastructure or even the whole system (for example, the country). It is recommended to distinguish CIs from critical systems. The latter has a broader scope and includes not only that CIs or CIs group, but also their communication, coordination, control and command units, in general, C4 infrastructure. It is clear that the C4 infrastructure, in turn, is also critical. This infrastructure may not be critical at first, but over time it will become critical by external influences or internal dynamics. For example, the loss of a segment of atomic (energy) technology which can be due to this or that external influence or internal decision-making and that processes can not only increase the vulnerability of the energy system and make it critical but also lead to the loss of stability and viability of the whole socio-economic system. Another example is that the loss (or under-development) of satellite capabilities can lead to blindness and, consequently, increase vulnerability to military and infrastructure development (Noravank Foundation, 2018: 163–164). There are many such examples in the recent history of both Armenia and many countries in our region.

The state policy development and implementation are in a quite important period in the RA, as Armenia has been in a new political phase since 2018. During that period deep reforms and transformations are taking place, in the field of both conceptual documents and normative-legal acts. Previous *National Security Strategy of the Republic of Armenia* defines that ensuring the reliability, security and safety of energy, transport, and communication infrastructure of the RA is a fundamental value of the national security of the RA, the disruption and disintegration of transit infrastructures to the spillover of ongoing military actions from neighboring states create a more diverse set of security threats for Armenia (NSS was approved at the session of the National Security Council at the RA President office on January 26, 2007). But interagency commission coordinating the process of developing the NSS was established, the personal composition of the commission and the order of activities were approved by the decree of the Prime Minister of the Republic of Armenia

Nikol Pashinyan. According to the decree, the Secretary of the Security Council was tasked to submit the NSS draft to the discussion of the Security Council of the RA. On July 10, 2020, the Security Council of the RA held a regular meeting in Yerevan, chaired by Prime Minister Nikol Pashinyan, and Armenia's NSS was approved (NSS was approved at the session of the Security Council on July 10, 2020). In this document protection of CI defines in these parts: "deterrence and neutralization of military threats" and "ensuring open and safe information" and "cyber domains".

In the normative-legal documents of Armenia, it is used the concept of objects of strategic importance, which in turn distinguishes and defines the concept of protection of special and objects of special importance. The protection of special and objects of special importance is carried out in accordance with both national legislation and international conventions.

15.2. Implementation

As already mentioned earlier, Armenia has undergone large-scale reform processes, in particular, an interagency commission has been set up to deal with defense reform, which is headed by the Secretary of the Security Council of the RA (Prime Minister of the Republic of Armenia, 2020). There is no unified body for the protection of special and important facilities in Armenia, as well as there is no coordinated policy, from which the actions of each structure can be derived. At present, each department implements the actions formulated within the framework of its powers and functions in that direction, which are enshrined in the charter or legislation regulating the field. In particular, it is necessary to touch upon several governing bodies: the Police of the RA, the National Security Service of the RA, the Ministry of Emergency Situations of the RA, the Ministry of High Technologies of the RA, the Ministry of Territorial Administration and Infrastructure of the RA, the Ministry of Defense of the RA. These bodies carry out their activities within the framework of the goals, objectives, functions and powers enshrined in the charter, and strategies and relevant legislation of their sphere.

The program schedule of measures to ensure the implementation of the provisions of the *National Strategy for Combating Terrorism in the Republic of Armenia* has been approved by the Government's decision of *November 22, 2012, N 1473-A*, which mentions several directions. They are:

1. Ensuring the unobstructed operation of strategic and engineering facilities of strategic and important objects.
2. Evaluation of the design threat of objects producing, transporting, using or storing radioactive materials and strengthening the physical defense system.

3. Establishment of information based on incidents related to the physical protection of nuclear energy use facilities, cases of detection of nuclear and radioactive materials.
4. Development of the capacity to eliminate terrorist acts, including cyberterrorism (President of the Republic of Armenia, 2012).

At the same time, the aforementioned structures are responsible for all these directions, which develop and implement the policy of their reserved spheres. On April 23, 2020, by the decision of the Government of the Republic of Armenia, the Reform *Strategy of the Police of the Republic of Armenia and the action plan for 2020–2022* based on that was approved, where in the demilitarization section of the Police Troops it is mentioned that one of the tasks of the Police Troops is to maintain special and important facilities and to ensure the safety of special cargo transportation (Government of the Republic of Armenia, 2020).

Until June 2018, the police forces of the RA mainly implemented functions of public order protection and public security. Police troops took part in part. Of the protection of the most important significant objects. With these changes, the National Guard as part of the Police of RA can, if necessary, protect the most important objects and government buildings. The lists of buildings and structures of state bodies and organizations subject to state protection under the Government of the RA, as well as of the most important objects, were approved in 1998 but were amended on July 20, 2004 (Government of the Republic of Armenia, 2004). These lists are also currently being changed. At the same time, the Division for Combating High-Tech Crime under the General Department on Combating Organized Crime at the Police of the RA is a centralized unit that is responsible for handling cybercrime with country-wide competency.

One of the tasks of the National Security Service (NSS) is to ensure the security of the military-industrial complex, nuclear energy, transport, communications, economy, finance and industry, strategic facilities, as well as in the field of primary scientific development. The function of the NSS is to implement state control in the field of information security in accordance with the law, as well as the coordination of the activities in the field of cryptographic and technical protection of the information protected by the legislation of the RA, the definition of a unified order counteracting technical intelligence organization in the RA (Prime Minister of the Republic of Armenia, 2018a).

Although Armenia has not published a formal cybersecurity strategy. Furthermore, a concept for an *Information Security Strategy* was developed in 2009, which outlines certain activities to be implemented. The *Digital Society 2020 Strategy* includes certain measures regarding cybersecurity activities and laws in the area of cybersecurity and data protection and is currently under discussion (Prime Minister of the Republic of Armenia, 2019). On October 23, 2017, the *Concept Paper of Information Security provision and information policy* was approved by the

President of the RA. In this document, there are provisions regarding information security, cybercrime and data protection. Cyber security is considered part of information security.

This document defines the concept of CII. Cyber-attacks on CII by individuals, organizations, and states are seen as a threat. One of the priority steps to ensure the security of cybersecurity and information infrastructure is to define and classify CII, create a registry, as well as the development of a legal act defining the relationship between the state and the holders of CII, the improvement the mechanisms for ensuring the security of CII in the RA.

One of Armenia's primary challenges in this area is the imperfection of the comprehensive state policy regulating the information and cybersecurity sector, the lack of legislation to protect CII, the insufficient level of institutional capacity of computer response agencies and the lack of a cyber security coordination structure. In order to increase resilience in the field of information, within the framework of the reforms adopted by Armenia, it envisages the development of national information and cyber capabilities through the formation of qualified professional potential and localized introduction of international standards.

One of the tasks of the Ministry of Emergency Situations is to coordinate the work implemented in the field of seismic risk reduction in the territory of the RA, to assess the seismic risk of the areas of special, important and general objects, which is fixed in 2018 by the approved charter (Prime Minister of the Republic of Armenia, 2018b). The Ministry of Emergency Situations carries out its activities by conducting a state examination of special, important objects, as well as objects, processes, projects and solutions that are dangerous in terms of civil defense, which is regulated by the RA *Law on Civil Protection*. In addition, this department has the function of developing medium-term programs for reducing seismic vulnerabilities of living systems and infrastructure of settlements, reservoirs, chemical and other objects of special importance, as it confirms the territory of the RA, seismic zoning and seismic risk assessment maps of its settlements, special, important areas of seismic protection in the field of seismic protection, expert seismic risk assessment. And the list of special, important and general objects in the field of seismic protection has been defined by the decision of the RA Government on March 13, 2003. The concept of the development of the seismic safety system of the RA was accepted in 2010 (Government of the Republic of Armenia, 2010). Until 2022, it is planned to establish civil defense bases in the border areas to ensure the protection of the infrastructure and infrastructure of the border areas.

The function of the Ministry of Territorial Administration and Infrastructure in the protection of critical infrastructure includes the development of normative legal acts in the field of energy, including the operation of energy facilities and safety rules and requirements, norms and restrictions on the use of land in energy zones, development and monitoring of safe operation and development of nuclear power plants, development and implementation of nuclear power plants, and one of the fixed tasks is the balanced and sustainable development of road, rail, water, air (except military) transport (hereinafter referred to as transport).

The Ministry of Transport, Communication and Information Technologies of the RA has been renamed the Ministry of High Technological Industry of the RA since June 1, 2010, according to the ministry, it has been given completely new functions, which means that there is a need to develop conceptual documents and programs regulating the sector. As for the RA Ministry of Defense, the functions of that department include the protection of critical infrastructure in the military sphere.

One of the most important issues for Armenia is the protection of critical transit infrastructure, which is regulated by international cooperation, as well as additional research, as it needs to be addressed. This issue is more than relevant for Armenia, as it has closed borders with two neighboring countries, namely Turkey and Azerbaijan.

Summarizing the information available on the protection of critical infrastructure in the RA allows us to fix the expediency of several actions. Thus

- To create an authorized body for the protection of critical infrastructure or a commission attached to the government, which will involve representatives of the bodies responsible for the protection of critical infrastructure, to develop a more systematic policy regulating the sector, inclusive norms and to regularly assess the situation,
- Develop a complex and comprehensive strategy that will clearly define the unified state policy for the protection of critical infrastructure, ensuring the security and resilience of those infrastructures in a coordinated manner,
- Implement support activities for more comprehensive and predictive research, monitoring and deepening and expansion of international cooperation, international cooperation,
- Provide appropriate action to develop the capacity of government agencies responsible for critical infrastructure protection to increase the level of security in a rapidly developing world.

15.3. Public-Private Partnership

Understanding that effective public-private cooperation, beyond strict adherence to legal requirements, is also a trust-building exercise extending beyond the letter of the law. Recognizing the need for informal but highly practical cooperation is of key importance.

Till 2018, the Police have provided protection in the private sector, too. The reason is that the special mechanisms are in transition and at that moment have no legal settlement for that. In the near future, the Police of Armenia will stop that practice and will continue to operate only in the areas which subject to police protection. This means that the RA pursues the development and application of mechanisms for the constructive cooperation between the public-private sector, as a type of the new culture. The processes taking place in the field of cyber security can be considered as the best practice in Armenia.

CERT.AM is the non-governmental academic CERT operating in Armenia. It is administered by a representative of the Armenian Internet domain (CERT-AM, 2018). The establishment of the CERT-AM was mandated via the Internet Society of Armenia decision on 1/09/2007.

AM NREN CSIRT is Armenia National Research and Education Network Computer Security Incident Response Team. It is administered by the representative of a major Armenian NREN which is ASNET-AM. CERT AM/AM NREN CSIRT is a national information security center operating under the management of the Internet Society of Armenia. CERT.AM/AM NREN CSIRT collects and analyzes computer incident cases (i.e. attempts or facts of violation of local rules and policies or rules globally accepted by Internet community on using computer resources), concerning network resources located in Armenia as well as responses to them with the aim of preventing, stopping and collecting evidence about an incident. At the moment, there are no private CERT/CSIRTs operating in Armenia. There is, however, cooperation with IT experts from private companies representing local and global/regional industries. Sometimes private sector experts provide expertise and advice to the Law Enforcement Agency officers, but in an informal format, based on personal relations and not within an established cooperation framework (Council of Europe, 2018: 21–26).

On November 23, 2015 the Investigative Committee signed a *Memorandum of Understanding* with ArmenTel, K-Telecom, UCom, Orange Armenia (these institutions belong to the telecommunication sector) with the intention of workload reduction and human resources saving. The main intentions of the Memorandum are to take effective joint measures in the direction of operative transmission of court decisions and transcripts and to develop mutual cooperation on the introduction of technical capabilities. In this framework, parties agreed to communicate in a standardized manner (including cover letters, electronic signatures) and agreed to cooperate in solving issues as soon as possible in case of such procedures. Furthermore, the signing providers agreed to process electronic transmission of the Investigative Committee within a short period of time and also to execute expeditiously court decisions that are designated as "urgent". A second objective of the Memorandum is to ensure that Internet access providers retain traffic data of their users for a period of time. Currently, Armenia has no data retention legislation (Council of Europe, 2017: 6–10).

In terms of other types of public-private dialogue, private sector stakeholders are invited on an occasional basis to the Parliament's relevant committees in order to provide their inputs. There is no established forum with regular meetings to exchange information on problems in cybersecurity. Instead, meetings are organized on a case-by-case basis.

There is a reported willingness of the government agencies to include the representatives of the private sector in the development of concepts, strategies and policies of the country, especially new *National Security Strategy* of the RA, which would be a welcome development in terms of public-private cooperation.

Thus, in the sphere of CIs protection, for the strengthening of public-private partnership contribution should include:

- Joint events,
- Development of mechanisms for the implementation of international and national standards for CIs by the private sector,
- multilateral cooperation and partnership.

15.4. Cooperation with European Commission

On January 2, 2015, Armenia joined the Eurasian Economic Union (EEU) (with Russia, Belarus, Kazakhstan and Kyrgyzstan). Despite Armenia's decision in September 2013 not to sign the Association Agreement with the EU, including a Deep and Comprehensive Free Trade Area Agreement (AA/DCFTA), Armenia and the EU continue their political and trade dialogue in areas where this is compatible with Armenia's new obligations to the EEU.

On October 12, 2015, the Foreign Affairs Council authorized the European Commission and the High Representative of the EU for Foreign Affairs and Security Policy/Vice-President of the European Commission to open negotiations on a new, legally binding and overarching agreement with Armenia, and adopted the corresponding negotiating mandate. The negotiations were officially launched on December 7, 2015, by the High Representative and the Armenian Foreign Minister. In March 2017, the negotiations were successfully concluded and the CEPA was initialed in Yerevan by both parties. On November 24, 2017, the Agreement was signed by the High Representative /Vice-President F. Mogherini and the Armenian Foreign Minister E. Nalbandian in the presence of the President of the European Council Donald Tusk and the President of the RA Serzh Sargsyan (Delegation of the European Union to Armenia, 2017).

Once ratified by all parties, the EU relations with Armenia will be based on the EU–Armenia CEPA, which was signed on November 24, 2017. The CEPA will provide for strengthening of cooperation in the areas of mutual interest, in particular, in political, economic and sectoral fields.

The CEPA provides a very solid basis to further enhance the reform agenda in the areas of mutual interest. The European Union and the RA have committed themselves to work together for the benefit of the citizens of Armenia and the

European Union, notably to contribute to the strengthening of democracy and of political, economic and institutional stability. Both sides committed to promoting, preserving and strengthening peace and stability at both regional and international levels and to enhancing cooperation in the area of freedom, security and justice with the aim of reinforcing the rule of law and respect for human rights and fundamental freedoms. This means that the CI protection is given an important role within the CEPA. The fifth chapter of the CEPA is devoted to the policies of the cooperation, especially transport, cooperation in the field of energy including nuclear security and environment.

Armenia was included (as a Southern Caucasus country) in the European Neighbourhood Policy (ENP) in 2004 and in the EaP in 2009.

In the framework of the revised ENP and the EaP, the new Agreement replaces the previous EU–Armenia PCA (European Communities and their Member States, and the Republic of Armenia, 1999), which provided for wide-ranging cooperation in the areas of political dialogue, trade, investment, economy, law-making and culture.

Summing up we argue Armenia will make efforts to construct deeper relations in some spheres with the EU for near future, including the critical infrastructure protection.

15.5. Education and Training

The critical infrastructure has important components such as scientific-educational and scientific research spheres. The main factor in ensuring the normal developing process of infrastructure protection is human with his intellectual/educational potential, as a collective capital. We get an opportunity to ensure full effective cooperation between the citizen and the state by developing the quality of the education.

Together training and education with internships are essential for succeeding in critical infrastructure security insurance and implementation of flexible and important programs: even for state officials, infrastructure owners and operators, first-person responsible and, of course, society, which strengthen the infrastructure and make the use of time and resources more rational.

Armenia is also guided by the aforementioned logic. They are various exercises that are carried out practically, including discussion and/or modeling of undesirable types of events and which may refer to critical infrastructure. We have many other types of preparation, which are based on constructive discussions between responsible operators of various agencies and infrastructures and contribute to the discussing, modeling, monitoring of specific scenarios and developing more profitable solutions for the resolution process: development of special plans, formation of a professional rapid response team, etc.

Thus, we should note that citizens' education, awareness, State-society relations and vice-versa cooperation are of paramount importance. Citizens' education is provided by several main professional institutions: Yerevan State University, National Polytechnic University of Armenia, and Crisis Management State Academy, which have relevant departments of different professions. In addition, civil society organizations are engaged in non-formal education. However, in order to ensure a high level of protection of critical infrastructure, it is necessary to constantly improve professional abilities and skills. Cooperation with the EU has great importance in this regard, within the framework of which one can get acquainted with the European methodology, models, and standards for crisis situations, to form a common approach.

Development and strengthening of defense means and mechanisms are of vital importance for Armenia, for which regular investments, educational sector reforms, measures developed with special priorities, modern technical systems in line with the development of the world are required.

Chapter Conclusion

Despite the RA having a dynamic development and an improvement in the protection of critical infrastructure and strengthening of resilience, there is a need to improve the policy in that sphere. Given the evolving dynamics of the relations between Armenia and the EU, the potential for progress and the new institutional level, it can be stated that critical infrastructure protection in Armenia will be boosted, after the ratification and the implementation stage of the CEPA Agreement, which will be based on cooperation with the European Union in several directions such as ideological, political, organizational and implementation should be.

Chapter 16

Azerbaijan

Alikhayyam Guluzade and Ayshan Suleymanova

Contents

This chapter provides an insight into how risks are assessed and managed for CII in general conditions, and it helps to understand how the government maintains its policies and programs for the protection of CII till today in Azerbaijan.

16.1. Introduction

In the transition to a global information society, states, societies, businesses and individuals face critical challenges in cyberspace in terms of the authenticity of information and its sources, secure use of e-services, protection of personal data, data integrity and confidentiality. In an environment where new cyber threats are constantly emerging and evolving, it is important for countries to have flexible, operational cyber security strategies against global cyber threats (İmamverdiyev, 2013).

DOI: 10.4324/9781003273769-19

All advanced countries have established a single governing body – the National Critical Infrastructure System (NISS) authority or the Critical Infrastructure Protection System (CRIS) authority, and national strategies have been adopted in this area. Its main objectives are to identify strategic facilities to be protected, increase resilience, ensure uninterrupted operation, analyze and prevent crises and risks, minimize losses in the event of an accident, and provide cross-sectoral coordination and monitoring to restore results as soon as possible.

The term "critical infrastructure" was first used in the US in the document *Executive Order (EO) 13010 Critical Infrastructure Protection,* established by a decree signed by the President of the United States in October 1996. *The Patriot Act* (2001), adopted after the September 11 attacks in the US, defines critical infrastructure (CI) as follows: a set of physical and virtual systems and tools that can lead to serious consequences (United States Congress, 2001: 131).

At present, a CI protection system has been established in most countries to regulate a group of vital facilities of strategic importance. Despite the global economic crisis, the CIP market is booming and will reach $135.48 billion in 2025 (Aliyev, 2020).

The CI systems in the Caucasus and Central Asia – the Greater Caspian Region – are of international importance with the reunification of both the Silk Road and the countries of the region. There is growing interest in this part of the world in terms of China's Belt and Road Initiative and its links to Afghanistan for military and civilian traffic. The South Caucasus and Central Asia also play a key role in China's new pipeline and road strategy, and Beijing is already successfully penetrating Central Asia and the Caucasus with promising economic projects, and the Caspian Sea region is an integral part of China's mega-projects. In other words, China has already joined the geopolitical confrontation and has become a third party. The geopolitical conflict in our region is no longer just between the West and Russia; the third player is China. Therefore, it is a crucial moment for all regional states, including Azerbaijan, to make a strategic choice (Muxtarli, 2018).

Currently, the four sectors – electricity networks, IT and telecommunication systems, the region's oil and gas pipelines, and new railways, ferries and other surface transport networks – also highlight how interconnected CI systems are; an accident in any of the previously mentioned damages the integrity and operations of others. As the President of the Republic of Azerbaijan, Ilham Aliyev noted,

> Although the causes of the accident are important, the most important issue is the breakdown of the overall system. A system must be set up so that the country's energy system does not collapse due to an accident. . . . An additional action plan must be prepared so that it does not affect the overall system in the future. This should be the main direction.

In light of the aforementioned, CI and its systematic management should be one of the most widely discussed topics on the agenda today. Although Azerbaijan shows more emphasis on management processes and systems on CII, in reality, the term *critical infrastructure* does not yet exist as a concept in Azerbaijan. This gives grounds to say that, in general, this field has not been mastered, whether in the scientific-theoretical, practical or normative-legal plane, and remains unaddressed by national policies or regulations.

16.2. Implementation

The idea of a "boundary or limit which should not be crossed", as highlighted in the speeches of the President of Azerbaijan at a meeting on 3 July 2018 regarding the accident at one of the substations of the Azerbaijan Thermal Power Plant – Mingachevir TPP. In this specific case, it was stressed that the coordinated activities and protection of these areas is the main direction to follow in order to guarantee adequate protection of vital infrastructures.

At the international energy forum held in Baku on 28 October 2020, the Azerbaijani Energy Minister said, "Ensuring flexible, transparent and high-quality governance in the energy sector in line with global challenges is a strategic goal of the Azerbaijani state, and all state support is aimed at modernization and effective regulation." Also, the world experience can be applied as a "mental model" for the consequences of the accident at Mingachevir TPP and the measures to be taken.

On 27 July 2020, the Government of Azerbaijan announced that it would restrict access to the Internet throughout the country, alongside the eruption of intense fighting between Armenia and Azerbaijan over competing claims to the Nagorno-Karabakh border region and its seven adjacent territories. Social media and communications platforms like Facebook, WhatsApp, YouTube, Instagram, TikTok, LinkedIn, Twitter, Zoom, Skype and Messenger servers were unavailable during such crisis. The restrictions came into place as a security measure declared by the Ministry of Transport, Communications and Technology to "prevent large-scale provocations from Armenia" as Azerbaijan declares martial law. Both countries have mobilized armed forces.

The largest oilfield in the Azerbaijan sector of the Caspian Sea is the Azeri-Chirag-Deepwater Gunashli oilfield (ACG). Production started in 1997. In 2011 production from ACG was 261.9 million barrels (an average of 717,600 barrels per day). In addition, about 3.3 billion standard cubic meters of associated gas was also produced in 2011. In total, ACG produced 1.9 billion barrels of oil from 1997 to the end of 2011.

The Shah Deniz gas field (SD), the largest gas field in Azerbaijan, was discovered in 1999. SD is geologically challenging and highly pressured. It has a reservoir thickness of more than 1,000 meters and is 22 km long. The field lies beneath water depths ranging from 50 to 600 meters. Production began in 2006. In 2011, SD produced about 6.67 billion standard cubic meters of gas and around 14 million barrels of condensate from four wells. Since the start of SD production in late 2006 till the end of 2011, about 30 billion standard cubic meters of gas and 63.7 million barrels of condensate were exported to the markets (BP in Azerbaijan Sustainability Report, 2011).

With the aforementioned energy infrastructure and growing energy potential, Azerbaijan, and subsequently the region, is playing an increasingly significant role in energy supply to the EuroAtlantic region. Today Azerbaijan produces 1.3% of the global oil (BP Statistical Review of World Energy, 2011).

Observers generally agree that the threat of a terrorist attack targeting energy infrastructure is real. Energy infrastructures are an attractive target for terrorists. Several groups have already indicated their intention and demonstrated their capability to conduct such attacks. Some major terrorist figures have clearly expressed their intention to strike global energy markets. In an audio message from December 2004, the Al Qaeda leader had explicitly called for attacks in the Gulf region and in the Caspian Sea (NATO Parliamentary Assembly document – Energy Security: Co-operating to Enhance the Protection of Critical Energy Infrastructures, 2008).

NATO could be important in the protection of energy resources and their transportation routes in the region. Increased level of NATO-Azerbaijan relations from cooperation in the frame of the Partnership for Peace (PfP) program to the execution of Individual Security Agreement would be in the mutual interest of the parties, as well as in the interest of energy consumers in the EuroAtlantic region.

However, new CI systems may already have weaknesses. For example, there are vulnerabilities in IT and telecom systems that are important for modern life and economies. As geopolitical tensions with neighboring countries increase in Azerbaijan, this undoubtedly results in the presence and increase of dangerous factors of special interest to various Azerbaijani government agencies and national security.

Cisco Talos researchers first discovered PoetRAT malware in April (April 17, 2020) in attacks on energy companies in Azerbaijan, including post-exploitation tools to log keystrokes, record images from webcams and steal browser credentials. The malware targeted software operators as well as SCADA, ICS and other victims in the public and private sectors of Azerbaijan. The malware gets its name from the references to sonnets included in macros included in malicious Word documents as part of a recent campaign by the first dramatic writer William Shakespeare. This time, the references in the macros – from the novel "Brothers Karamazov" by Russian writer Fyodor Dostoevsky – can also be a covert reference to the current conflict. Both Azerbaijan and Armenia were formerly part of the former Soviet Union, and Russia has close ties with both countries and is also Armenia's military ally (Montalbano, 2020).

16.3. Public-Private Partnership

Country profiles on European Small Business Act (ESBA) assessment "SME Policy Index: Eastern Partnership 2020" is part of the publication. The country assessment documents comprehensively analyze the implementation of the 10 principles of the ESBA in the six EaP countries (Armenia, Azerbaijan, Belarus, Georgia, the Republic of Moldova and Ukraine). Based on the opportunities for improvement, the problems faced and remaining by the countries are identified. After a period of strong growth in 2001–14, Azerbaijan's GDP fell in 2016–17, as a result of falling oil prices, along with the economic slowdown in regional trading partners. The recovery period began in mid-2017, with an increase of 1.4% the following year, supported by rising oil prices and more favorable economic conditions in the region, leading to a sharp rise in inflation in 2016 and 2017 (12.4% and 12.8%) economic instability and the depreciation of the local currency led to a banking crisis. Twelve banks exited the market, and lending to the private sector fell from 38.5% of GDP in 2015 to 20.8% in 2018, below the average levels of EaP and OECD of 38.3% and 93.9%, respectively (World Economic Forum, 2019).

The amount of foreign direct investment (FDI) has increased over the last two decades, reaching 70% of GDP in 2017 (UNCTAD, 2018). The oil and gas sector receives the most foreign investment, while other sectors received only 10% of the investment. The oil and gas sector accounted for 38% of GDP in 2018, and crude oil, petroleum gas and refined oil (The State Statistical Committee of the Republic of Azerbaijan, 2019). Thus, diversification of economic activity, employment and exports has become an important policy priority. In terms of agricultural employment, it is the largest sector (34% of the total active population and 5.6% of GDP in 2017), but the sector suffers from low productivity and investment due to fragmentation. In 2017, production was only 5.2% of total employment – and 4.7% of GDP and is being further developed (The State Statistical Committee of the Republic of Azerbaijan, 2019).

To achieve the goals outlined in the SME roadmap, the government has stepped up administrative requirements, expanded e-government services and implemented a number of policy initiatives aimed at promoting the private sector. As a result, the country ranked 34th in the 2020 World Bank Doing Business ranking. The most notable developments were in the areas of credit availability and protection of small investors (World Economic Forum, 2019). Azerbaijan ranks 58th out of 141 countries analyzed by the Global Competitiveness Index (GCI) of the 2018 World Economic Forum (ibid, 2019).

The Ministry of Economy of Azerbaijan has launched a new initiative to finance small and medium enterprises. Funding will be provided by KOBIA, and small and medium Business Development Agencies. The project is designed to promote startups, SME training, international exchange programs and technological innovation.

At present, the policy in this area is aimed at ensuring the implementation of state support for entrepreneurship in the form of a system, increasing the effectiveness of state support measures for entrepreneurs. Consistent measures are being taken to implement this strategic line. A mechanism of state financial support for entrepreneurship has been established, which serves as an important real source of meeting the financial needs of small and medium entrepreneurs.

During the NFES, more than 2.1 billion manat[1] of soft loans were provided to more than 35,300 entrepreneurs to finance investment projects worth 4.5 billion manat. 75% of the loans fall to the regions of the country, and 25% to Baku settlements. 90% of the financed projects are small loans. One of the main activities of the NFES is to educate entrepreneurs. So far, the foundation has held 362 business forums with the participation of more than 48,000 entrepreneurs. Regular measures are taken to strengthen business consulting, information provision and development of business relations (The Igtisadiyyat, 2018).

16.4. Cooperation with European Commission

The EU and Azerbaijan have worked together to conclude various political agreements. The most important is the EU-Azerbaijan PCA (1999). Azerbaijan is part of the European Neighborhood Policy (ENP) and the EaP initiative. Relations with Azerbaijan are defined in the context of the EaP around two key principles: 1. the need for a differentiated approach; Azerbaijan is free to choose to which extent it will participate in the EU's offer of political association and economic integration; 2. the implementation of deep democracy: the EU will not compromise on the core values of the EaP (GDSI Limited, 2017).

EU support to Azerbaijan amounts to around €30 million each year. Non-government organization (NGO) projects under Human Rights, Democratization and Non-State Actor budget lines (EIDHR and NSA) also get funding. Civil society in Azerbaijan is supported through the European Instrument for Democracy and Human Rights (EIDHR) and the Non-State Actors Programme (ibid.).

Azerbaijan economy is based on oil and gas extraction which account for 95% of exports and 75% of government revenues. Azerbaijan has an average number of SMEs per capita in relation to the other Eastern Partner countries, but certainly the lowest share of employment generated by the SME sector. Likewise, their share in GDP represents just a small fraction of national economic output, otherwise dominated by the oil sector.

Recent data indicates that stumbling activity in H1 due to the pandemic sapping global oil demand and prices, carried over into Q3. GDP contracted 3.9% on an annual basis in January–September (January–June: –2.7% year-over-year) as both the hydrocarbon and non-hydrocarbon segments of the economy continued to shrink. Similarly, the decline in industrial output sharpened through September as a reeling mining sector, dragged down primarily by falling oil production, outweighed solid gains in the non-oil and gas sector. Meanwhile, on 24 October 2020, S&P Global Ratings affirmed the country's credit rating at BB+ but downgraded the outlook to negative, due to rising risks to its macro stability and fiscal and external positions amid the military confrontation with Armenia. The conflict over the disputed Nagorno-Karabakh region shows little sign of abating, with various attempts at internationally brokered ceasefires failing.

The economy is seen suffering its sharpest downturn since the late 1990s this year on the impact of the pandemic and battered energy industry. The output should bounce back to growth next year, however, driven by a recovery in external demand and hydrocarbon prices. The recent surge in COVID-19 cases and the escalating conflict with Armenia pose key risks to the outlook. Focus Economics panelists see GDP growing 2.9% in 2021, which is up 0.1 percentage points from last month's forecast, and 2.6% in 2022 (Focus Economics, 2020).

Azerbaijan has a workforce of 5.07 million out of its 10 million population (World Bank data). Its economy is based on gas and oil, steel, iron, chemical and petrochemical products, and textiles. Agriculture accounted for 5.2% of GDP and employed 35.8% of the population (World Bank). Main crops include wheat, barley, corn, fruits (wine grapes), potatoes, cotton, tea, silk and tobacco. The country also produces other potentially valuable crops, including indigenous pink grapes and persimmon. Soon, a new online and mobile government platform will allow Azerbaijani farmers to sell their products.

Industry accounts for 52.2% of GDP and employs 14.4% of the population (World Bank). Besides oil products and their derivates, Azerbaijan produces cement, machinery, cotton and foodstuffs. In the large-scale Azerbaijan – 2020: outlook for the future development project, the Azerbaijani Government seeks to more than double the volume of per capita GDP in the country through non-oil industries.

Services account for 35.2% of GDP and employ 49.6% of the population. Flourishing service sectors include banking, construction and real estate. Situated in 9 out of 11 existing climatic zones, Azerbaijan has quite a unique geographical position and privileged tourism potential. 2.9 million tourists visited the country in 2019 (State Border Service), a growth of 11% compared to 2018. Widespread salt springs in the country are progressively turned into flourishing health spa resorts. Furthermore, Azerbaijan and Uzbekistan reached an agreement to jointly prepare tourist guides with international status (Nordea, 2020).

There have been faster growth in imports from the EU than exports to it, although trade in both directions has been on the rise. Azerbaijan currently imports EUR 2.7 billion and exports 11.5 billion worth of goods and services from and to the EU respectively, making it the only EaP country to have a trade surplus with the EU (EU4Business, 2019).

16.5. Education and Training

The science and education system should be geared toward research and development on the subject of CI protection. It should be added to the relevant curricula in the education system, and classes and training should be provided. Against the background of increasing frequency of crises and threats, the inadequacy of society's attitude to them is called the "sensitivity paradox". We would like to note that the leadership of the Center for Strategic Studies (CSS) under the President of the Republic of Azerbaijan recommended to take into account the urgency of this topic during presentations in various universities and institutions of the country, expressed support for the development of relevant scientific and theoretical materials.

When discussing emerging security challenges, the protection of CEI occupies an important place. The 2010 Strategic Concept commits NATO not only "to enhance the capacity to detect and defend against international terrorism" but also to "develop the capacity to contribute to energy security, including protection of CI and transit areas and lines". These commitments entail analysis of the threat, strategic assessments, consultation, planning and cooperation with partners. Cooperative efforts by the NATO Allied/Partner community in the area of overlap between counterterrorism and protection of energy infrastructure constitute an important response to these requirements (Mission of the Republic of Azerbaijan to NATO, 2020).

Mindful of this, NATO considers Azerbaijan an important ally in cooperation on energy security, and Azerbaijan actively participates in EAPC activities on energy infrastructure protection and countering terrorist threats in this field. Since March 2008 Azerbaijan has chaired the informal EAPC PAP-T Working Group on the Protection of Energy infrastructure, in the light of its vast experience in energy security. The aim of this group is to develop a common understanding of the security threats and risks to energy infrastructure. A shared understanding of these challenges is essential to enhance ideas and best practices on counterterrorism and other threats to energy infrastructure. Azerbaijan and NATO are continuing a political dialogue and exchanging experience on the protection of CEI as well (ibid.).

Azerbaijan attaches importance to cooperation with NATO in the field of anti-terrorism, which the international community must tackle together. NATO has been actively engaged in the fight against terrorism and focuses on improved threat awareness and preparedness, developing adequate capabilities and enhancing engagement with partner countries and other international actors. Azerbaijan is building its interaction with the NATO and EAPC countries in a wide range of areas relevant to the fight against terrorism, including the military response to terrorism, enhancement of national counterterrorism training capabilities, intelligence and information sharing, border security and civil protection anti-terrorist measures.

Finally, it should be noted that currently in our country, CI facilities and subjects (operators) operate in isolation from each other. That is, the more CI objects there are, the more management subjects there are and each of them operates independently. This complicates this interaction, monitoring the situation on a 24/7 basis, analysis to identify "weak points" and the continuous exchange of information, and flexible regulation in the face of potential risks.

Chapter Conclusion

In this regard, the first step is to develop a relevant concept in the field of CI protection. In addition, existing laws should be examined to take into account crimes that occur in specific circumstances. Today, there is no normative legislation in Azerbaijan that defines who and how protects the inter-sectoral activities of the CI from potential risks and threats. Legal regulation in the field of information security in our country remains unsystematic and uncoordinated. The existing normative-legal acts are mainly aimed at regulation in the field and cover the state sectoral regulation. The first step in developing and implementing a management (regulatory) policy adequate to global challenges may be to define a national action plan. Here you can find the main goals, objectives and tasks.

Protecting a country's CI is important for its security, economic well-being and the well-being of its people. Protecting critical infrastructure is as important as the construction of these systems, but often the protective/restorative work on these systems is delayed, not funded or taken into account. The consequences can range from anxiety to death, from public discontent to serious political violations. Companies and governments must play a role in identifying threats and needs in the design and installation of security solutions.

The formats of international cooperation created in the field of CI should be applied in Azerbaijan as well. For example, in 2012, the US, the UK, Canada, Australia and New Zealand signed the Critical-5 agreement. From this point of view, "Baku-Tbilisi-Ceyhan Oil Pipeline, Baku-Tbilisi-Erzurum Gas Pipeline and Baku-Tbilisi-Kars railway", etc. are objects of CI and their protection is an integral part of the energy security of the region (South Caucasus and Europe). According to the aforementioned model, Azerbaijan also has a common CI for the energy security of Europe and the countries of the region: "Critical-3" or "5", etc., which provides joint protection of the pipeline and logistics network can sign format agreements. Cooperation with the same content can be concluded between our country and the Central Asian republics, which have become a logistics center, a transport hub of Eurasia.

As a result of the successful implementation of the e-government and digitalization policy pursued in the country, the CI sectors as a whole are related to this area. Each government agency is increasing the number and types of e-services. The establishment of a number of agencies related to information security and energy regulation in the course of recent reforms, and the coordinated activities of these agencies also indicate the formation of an important basis for the establishment of a single ICRC in the country in accordance with world standards.

The establishment of the CI protection system is of strategic importance in further strengthening the statehood, ensuring stability and a more prosperous life of our people in Azerbaijan, which is currently one of the most stable places in the world.

Note

1. The Azerbaijan Manat is the currency of Azerbaijan.

TRANSATLANTIC COOPERATION

4

United States of America
Canada

DOI: 10.4324/9781003273769-20

Chapter 17

United States of America

Matthew Vatter and Richard J. Larkin

Contents

This chapter gives an insight into the system for CIP in the US. It discusses the development of the system, modern frameworks under which CI stakeholders and owners address threats, prevention, protection, detection, response, and recovery programs and the documents that provide guidance to ensure continuity and agreed-upon standards. The chapter discusses public-private partnerships and international dependencies and collaborative efforts. The chapter concludes with the authors' thoughts on the way forward for CIP programs for the US and the necessity for continued collaboration across international borders.

DOI: 10.4324/9781003273769-21

17.1. Introduction

The US government's efforts to develop a comprehensive national policy for CI began in the mid-1990s. *Executive Order (EO) 13010 Critical Infrastructure Protection,* 1996 identifies "certain national infrastructures so vital that their incapacity or destruction would have a debilitating impact on the defense or economic security of the United States." Threats are primarily determined as physical and cyber (The White House, 1996)[1]. The EO also established bodies and mechanisms with the aim of developing a normative framework for CIP in the US. In 1996, President Clinton established the President's Commission on Critical Infrastructure Protection (PCCIP), with a charter to designate CI, to assess their vulnerabilities, to recommend a comprehensive national policy and implementation strategy for protecting those infrastructures from physical and cyber threats, and to propose statutory or regulatory actions to affect the recommended remedies.[2] CI was defined in 1998 *Presidential Decision Directive/NSC-63 Critical Infrastructure Protection* as

> physical and cyber-based systems essential to the minimum operations of the economy and government. They include, but are not limited to, telecommunications, energy, banking and finance, transportation, water systems and emergency services, both governmental and private.
>
> (The White House, 1998)

Further, it is emphasized how

> [m]any of the nation's critical infrastructures have historically been physically and logically separate systems that had little interdependence. As a result of advances in information technology and the necessity of improved efficiency, however, these infrastructures have become increasingly automated and interlinked. These same advances have created new vulnerabilities to equipment failure, human error, weather and other natural causes, and physical and cyber-attacks. Addressing these vulnerabilities will necessarily require flexible, evolutionary approaches that span both the public and private sectors, and protect both domestic and international security.
>
> (ibid., 1998)

17.2. Implementation

The US identifies 16 CI sectors. The systems and assets in these sectors are vital to the structure and operations of the national economy, public health and safety, and the overall functioning of American society. *Homeland Security Presidential*

Directive 7 created a national policy for federal departments and agencies to identify and prioritize critical infrastructure (Department of Homeland Security, 2003). *Presidential Policy Directive 21* (PPD-21) outlines the policy enabling a strong, resilient and secure system for protecting American critical infrastructure (The White House, 2013). Having federal-level guidelines to direct efforts of federal agencies establishes roles and responsibilities across government and creates the foundation upon which protection activities occur.

Separating CI into 16 sectors facilitates the assignment of sectoral responsibilities within the government and to private industry stakeholders. SSAs are identified to provide a lead resource for the organization of multi-agency and stakeholder efforts to secure key sector assets. SSAs, in coordination with the Secretary of Homeland Security, prioritize CI based on threat and vulnerability analysis, collaborate with sector-specific CI owners and operators, carry out incident management, provide technical support and assistance, and help mitigate incidents. SSAs are also responsible for regular reporting to the DHS the overall state of preparedness within their assigned sectors and identifying areas of concern. SSAs are charged with considering CIP from an "All-Hazards" approach, which means that incidents and threats considered come from natural *and* human-caused sources. All hazards include natural disasters, industrial accidents, acts of terror, pandemics, cyber incidents, sabotage, and destructive criminal activities that target CI.

Table 17.1 identifies the 16 CI sectors and the SSAs for each sector as well as a brief description of the areas of responsibility for each sector:

Threats to the nation's CI under this approach fall into five categories: direct enemy attack, cyber threats, accidental or technical failures, extreme weather, and pandemics (Department of Homeland Security, 2013a: 8). A particularly important lifeline sector is energy. The importance of electricity in daily life is obvious and often taken for granted. Events or actions in any single threat category could result in significant disruption to the electricity distribution system. Combined attacks in multiple threat categories could result in catastrophic, long-term loss of electric power and disrupt sectors dependent upon it

Exacerbating the vulnerability of America's critical infrastructure is the fact that no single organization has authority over it. No matter how diligent CI owners are, the achievement of a cohesive, synchronized effort to protect the vast number of interdependent portions will likely not be fully achieved. It is estimated that approximately 85% of CI is privately owned (Government Accountability Office, 2006). Protecting the nation's infrastructure to include the electric grid requires cooperation and communication among all the stakeholders, governments, private industry, and local communities. In the next section, we will discuss the relationship between government at the federal, state, local, tribal, and territorial levels and the private sector owners of CI.

Table 17.1 US Critical Infrastructure Sectors and Sector-Specific Agencies

Sector and Specific Agency	*Description*
Chemical Department of Homeland Security	The chemical sector is responsible for the manufacture, storage, use, and transport of potentially dangerous chemicals relied upon by a wide array of other critical infrastructure sectors.
Commercial Facilities Department of Homeland Security	The commercial facilities sector includes sites that draw large groups of people for lodging, business, entertainment, and shopping. These facilities are characterized by having open access to the general public with a vast majority being privately owned.
Communications Department of Homeland Security	The communications sector provides an enabling function across all sectors of critical infrastructure. It includes terrestrial, satellite, and wireless communications.
Critical Manufacturing Department of Homeland Security	The critical manufacturing sector includes those manufacturing capabilities that underpin many portions of other critical infrastructure sectors. They include primary metals manufacturing, machinery manufacturing, electrical equipment, appliance and component manufacturing, and transportation manufacturing.
Dams Department of Homeland Security	The dams sector provides critical water retention and control which includes hydroelectric power generation, agricultural irrigation, sediment and flood control, river navigation, and industrial waste management.
Defense Industrial Base Department of Defense	The defense industrial base sector is comprised of domestic and foreign companies that provide the materials and services necessary to build, maintain, mobilize, deploy, and sustain US military operations.
Emergency Services Department of Homeland Security	The emergency services sector provides day-to-day as well as emergency response and recovery services. It is organized primarily at the federal, state, local tribal, and territorial levels of government and includes police, fire, and emergency medical services organizations capable of all-hazard response.
Energy Department of Energy	The energy sector describes the infrastructure that provides energy resources underpinning all sectors of critical infrastructure. 80% of the country's energy infrastructure is privately owned.

Financial Services Department of Treasury	The financial services sector includes depository institutions, investment institutions, insurance companies, and other credit and financing organizations that enable the transfer of financial products and services.
Food and Agriculture US Department of Agriculture and Department of Health and Human Services	The food and agriculture sector provides for the production, manufacturing, and storage of the nation's food and agricultural products supply.
Government Facilities Department of Homeland Security and General Services Administration	The government facilities sector oversees buildings located in the US and overseas owned or leased by the US government that house embassies, military installations, courthouses, national laboratories, critical equipment, and systems.
Healthcare and Public Health Department of Health and Human Services	The healthcare and public health sector protects all sectors from terrorism, infectious disease, and natural disasters. It is responsible for prevention, resiliency, response, and recovery to public health-related issues.
Information Technology Department of Homeland Security	The information technology sector encompasses the people, hardware, and software necessary for the function of government, academia, private sector businesses, and the general public. In coordination with the communications sector, is responsible for the internet.
Nuclear Reactors, Materials, and Waste Department of Homeland Security	The nuclear reactors, materials, and waste sector includes nuclear power generation, medical isotopes, and nuclear and radiological research. It also oversees the movement of radiological cargo in coordination with the transportation sector.
Transportation Systems Department of Homeland Security and Department of Transportation	The transportation sector moves people and goods. It includes aviation, highway and motorway, maritime, mass transit and passenger rail, pipeline systems, freight rail, and postal and shipping.
Water and Wastewater Systems Environmental Protection Agency	The water and wastewater systems sector is responsible for the nation's clean water supply. It also is critical in the management of sewage and wastewater treatment.

Source: Derived from NIPP 2013

17.3. Public-Private Partnership

The protection of CI in the US is a shared responsibility between private sector owners in each of the sectors and governmental agencies at the federal, state, local, tribal, and territorial levels. The US government, in coordination and collaboration with key stakeholders, has developed a number of standardization documents that provide structure to protection programs across the breadth of the 16 sectors.

The government and the private sector each have unique responsibilities and the perspectives of each are equally important. Plans are maintained through the use of Sector Coordinating Councils, Government Coordinating Councils, and Regional Consortia. These collaborative bodies are further broken down to focus on specific areas within a sector. For example, in the financial services sector, the Financial Services – Information Sharing and Analysis Center (FS-ISAC) acts as the financial industry resource for cyber and physical threat intelligence analysis and sharing. This is a membership-based organization comprising the private sector, government, non-profit, and select partner stakeholder organizations. The National Council of ISACs is a coordinating body that facilitates the collaboration of sector-based ISACs. ISAC members contribute sector-specific information and intelligence in an effort to share knowledge and experience. This voluntary collaboration among sector stakeholders is a key element in the overall *National Critical Infrastructure Protection Plan* and is vital to successful CIP efforts. As part of the National Infrastructure Protection Plan (NIPP), public and private sector partners in each of the 16 CI sectors and the state, local, tribal, and territorial government community have developed Sector-specific plans that focus on the unique operating conditions and the risk landscape within each unique but interdependent sector. In 2015 sector-specific updates addressed the nexus between cyber and physical security, interdependence of sectors, risks associated with aging infrastructure, outdated technology and climate change, and changes in the workforce needed to continue to support the national plan. At the beginning of 2020, there have been no updates published for sector-specific plans.

The proliferation of cyber-attacks necessitated a new approach toward synchronizing action, collaborating on public-private efforts, and coordinating intelligence. In 2018, to better address the growing complexity of protecting the nation's CI and to centralize historically decentralized efforts, the *Cybersecurity and Infrastructure Security Agency Act* created the CISA. The CISA is a result of the collaborative effort to identify gaps in communication and coordination and develop a method to better address areas of operational weakness. It is an example of how governments must continually assess structure and oversight and be willing to adapt and adjust to best meet changing operational and strategic environments.

CISA is the single coordinating element for public and private sector organizations. CISA operates the National Cybersecurity and Communications Integration Center (NCCIC), which provides real-time national-level awareness to cyber threats, coordinates national-level response in the event of large-scale cyber incidents, and facilitates cross-sector awareness to emerging cyber threats and attack methodology. CISA works with private sector CI owners and state and local governments on CI resiliency efforts. CISA's national focus allows for a central repository of best practices and sharing of resiliency resources and expertise. CISA conducts risk analysis and vulnerability assessments of the nation's CI. This analysis becomes the foundation for local, regional, and cross-border collaborative resilience projects.

17.4. Cooperation with European Commission

In 2011, the Department of Homeland Security, through the Federal Emergency Management Agency (FEMA), entered into a partnership agreement with the Directorate-General for Humanitarian Aid and Civil Protection of the European Commission. The general purpose of the agreement fell into three main areas: 1) to enhance disaster resilience, reduce disaster risk, and improve international response and coordination between the participants; 2) to explore the process and share best practices for facilitating coordination during disasters; and 3) to expand the respective knowledge bases concerning prevention (including adaptation to climate change), preparedness, response, and recovery and resiliency activities (DHS, FEMA, DG ECHO, 2011). This agreement focused primarily on administrative professional exchanges of best practices, lessons learned, the exploration of methodologies, and collaboration between the governmental entities in the US and Europe that primarily manage disaster preparedness and response. Participants would self-fund any events such as participation in conferences, training events, or observation opportunities. The US participation often includes EM practitioners, government officials, medical professionals, academia, and the military. FEMA also grants international students access to the Emergency Management Institute (EMI) and the National Fire Academy in Emmitsburg, Maryland, and the Center for Disaster Preparedness in Anniston, Alabama. Programs are offered in both resident and online formats. These programs offer practitioner, management, and leadership training tracks.

Cooperation often occurs at the sector level. The U.S. Department of Energy maintains scientific ties with the EU and the individual Member States. These relationships facilitate joint publication, industry expert exchanges, and joint workshops to mention a few areas of routine collaboration. In 2009 the EU and US formed the U.S.–EU Energy Council. The council is supported by three working groups: Energy Technology Group, Energy Policy Working Group, and the Energy Security Working Group. Among the stated goals of the council are "Promote security of transit and key energy infrastructure that could improve energy security and the regional and global level" and "Promote strengthened power grids to facilitate deployment of renewable and low-carbon sources of energy"(EU-U.S. Energy Council, 2009). The council has convened eight times since its inception. In

May 2019, the EU-US Energy Council sponsored a business forum that focused on liquefied natural gas (LNG) export and import. Providing a forum for relationship building, the forum brought the US and European companies together, facilitated discussions on import and distribution to EU members, and established pricing mechanisms (CEPCONSULT, 2019). The EU-US Energy Council continues to develop business relations and policy discussions to facilitate sustainable, reliant, and diversified energy resources for member nations at strategic as well as operational levels.

US–EU cooperation often occurs bilaterally between US entities and the partnered EU states. Most recently, through the State Partnership Program and the US Embassy in Croatia, the US committed funds and expertise, in the form of training and exercise exchange, to upgrade Croatian cyber defense capabilities (U.S. Embassy in Croatia, 2020). Initially focused on Ministry of Defense priorities, the capabilities developed will establish a foundation for greater cybersecurity capabilities across all ministries.

Events against CI do not happen in a vacuum. Effects cascade and result in threats and or disruption to other aspects of CI. A recent Government Accountability Office (GAO) report found that the proliferation of internet-connected devices and the increasing reliance on connectivity across all sectors has increased the vulnerability of CIP systems (US GAO Report, 2019). This report highlights the interconnected and interdependent nature of CI systems. Expanding this to international dependencies, it is obvious that the emerging global threat cannot be addressed by a lone-nation approach, where each country acts without regard to its neighbors. Actions against Ukrainian power production and distribution identified a threat vector by which any nation can be vulnerable. Sharing analysis of these events, including intelligence and mitigation, is the only way all stakeholders can quickly adapt. Moreover, ensuring that cross-border solutions mitigate cascading effects across sectors is essential. Programs such as RECIPE project, Erasmus, RACVIAC, and the State Partnership Program, focusing on CIP collaboration involving the US, Canada, the European Union, and NATO, facilitate the exchange of ideas and give practitioners opportunities to train together. This cooperation has worked to address general security concerns across the globe and is a key component in developing resilient CIP systems that transcend international boundaries.

17.5. Education and Training

Education and training in CIP in the US is extensive and widely available. The main source of education and training for CIP is the Department of Homeland Security, through a formalized delivery system, consisting of direct delivery and identified training and education providers, supported through DHS funding. A shortcoming of the training and education system for CIP in the US is a lack of a universally accepted curriculum and standards for CIP education and training.

There are recommended courses and for some higher education institutions, there are developed curriculum and syllabi to support the delivery of courses, but the content is based upon the author's interpretation of the elements of the NIPP, along with the sponsoring institution practices and policies for curriculum development.

Federal CIP Education and Training

The NIPP establishes a framework to build national awareness and enable the education and training needed to ensure an effective, efficient CI and key resources (CIKR) protection program (Department of Homeland Security, 2013b).

The focus of the NIPP framework is to build awareness for CIP and to help create expertise in the delivery of CIP in the field. Education and training opportunities are also delivered within and across the various SSAs, and other federal departments and agencies. The increased focus on cybersecurity is also evident in coordinating with the CIKR sectors on efforts to increase cyber security, such as DHS' Control Systems Security and Critical Infrastructure Protection Cyber Security Programs. These programs provide the sectors with subject matter expertise, recommended practices, tools, and training courses. In addition, DHS also conducts exercises to identify gaps and to validate plans and training. Cyber Storm is a congressionally mandated biannual exercise sponsored by DHS to engage public and private sector cybersecurity and infrastructure security partners to strengthen cyber preparedness (Department of Homeland Security, 2019).

As discussed in Section 17.3, public-private partnerships are core to the US CIP program. The North American Electric Reliability Corporation (NERC), a not-for-profit international regulatory authority, sponsors the Grid Security Exercise program, GridEx. This program is intended to provide realistic scenarios in which utilities can demonstrate their capacity to respond to complex, coordinated cyber and physical security threats and incidents. The participating audience is composed of private utility companies, state, local, provincial and federal governments, CI cross-sector partners such as ISACs and supply chain stakeholders. US military cyber and domestic response elements are also invited to participate in GridEx. This "whole of enterprise" approach uncovers gaps in communication, misunderstanding in decision-making authority, and other strategic issues that enable each entity to refine and adjust organizational plans. It also facilitates dialog between senior government officials and private sector utility executives.

Consortium

The US National Domestic Preparedness Consortium consists of seven entities providing delivery of training and education to public, private, and non-governmental personnel in furtherance of the National Preparedness Goal. There are over 225 individual courses available through the consortium members providing training and education in CIP, ranging from Cybersecurity Protection to Infrastructure Systems Protection. Courses are designed for three levels of competency: *awareness level* is designed for responders who require the skills necessary to recognize and report a potentially catastrophic incident or who are likely to witness or investigate an event involving the use of hazardous and/or explosive devices; *performance level* is designed for first responders who perform tasks during the initial response to a catastrophic event; *management and planning* courses are designed for managers who build plans and coordinate the response to a high-consequence event (either human-caused or natural hazard).

Colleges and Universities

The FEMA Higher Education Project offers many certificate and degree programs (undergraduate, masters, doctorate) (FEMA, 2020). Created in 1994, the vision of the project is to "Serve as the Nation's leading focal-point for EM higher education, foster the professionalization of the field via educational efforts, and contribute to a more resilient nation by creating a cadre of more professional emergency managers" (FEMA, 2014: 6). The FEMA Higher Education Program is arguably the most comprehensive source of information regarding EM and Homeland Security Education and Training in the United States.

Throughout the US, there are three main "tracks" for formal education – EM, Homeland Security, and International Disaster Relief/Humanitarian Assistance. Each of the three areas has education opportunities at the Certificate, Associates, Bachelors, Masters, and Doctorate levels.

State Delivery and Practitioner Certification

State of Minnesota Department of Public Safety, Division of Homeland Security and Emergency Management (HSEM) provides several professional certificate programs called learning paths. They are designed for learners who are currently working in EM or have a role in emergency preparedness in Minnesota. The curriculum for each learning path consists of a combination of FEMA online Independent Study (IS) courses and HSEM direct-delivery training classes and workshops (Minnesota Department of Public Safety, 2020).

Formal education programs, certificate programs, and practical training courses continue to emerge. The changing nature of CIP requires a flexible education and training approach. Increased dependency on digitally connected systems across all sectors requires a greater emphasis on cyber-related training. The way practitioners, technicians, supervisors, executives, and government officials are trained must continue to develop. Exercise programs such as GridEx, Cyber Guard, and Cyber Storm must be continued and expanded to address the changing nature of the threat landscape. These efforts are the foundation for coordination and communication across all sectors and among all stakeholders.

Chapter Conclusion

In the US, CIP is a fairly well-established program. However, the dynamic nature of threats, man-made and natural, necessitates agility, flexibility, and a willingness to adapt to those dynamic threats. Over time, the US learned that it must train practitioners at all levels of the government and within the private sector. CISA is an example of how the US federal government has adapted to the need for greater levels of coordinated CI training, exercise, and resiliency. The many programs identified in Section 17.5 will be better coordinated under this federal agency. By centralizing the activities required to protect CI, greater efficiency and effectiveness can be achieved. International training programs offered by FEMA and international programs that put practitioners and leaders from many nations together in a collaborative training environment address the interconnected nature of CIP systems and related vulnerabilities.

It will take the efforts of all concerned parties, government, private companies, non-profit organizations, and individual citizens and communities to solve the complex problems identified in this chapter and throughout this text. There is not a *right* answer or a *right* process. No one set of laws, policies, or organizational structure can be applied. The best ideas and an understanding of failures and triumphs must be understood and applied within the context of the physical and geo-political environments in order to mitigate the plethora of existing and emerging threats.

Notes

1. Executive Orders are official documents, through which the President of the United States manages the operations of the Federal Government.
2. The PCCIP submitted its report, Critical Foundations: Protecting America's Infrastructures, in October 1997. The PCCIP report was the basis for Presidential Decision Directive 63 (22 May 1998), Critical Infrastructure Protection, which establishes national policy and an organizational structure for effecting a public-private partnership and for accomplishing the special protection functions that are inherently the responsibility of government (The Department of Defense, 1998).

Chapter 18

Canada

Ali Sheraz

Contents

This chapter provides an insight into Canada's CIP's efforts in recent times. Starting from 2010, the Government of Canada has launched a set of initiatives in order to better define the CIP field and national frameworks. Such initiatives, together with more recent policies and strategies, have defined Canada's national governance in CIP and have consequently triggered the definition of competencies and responsibilities in this strategic field. The chapter also takes into account the US–Canada relationships and joint efforts in CIP.

DOI: 10.4324/9781003273769-22

18.1. Introduction

Like many advanced countries in Europe, Canada has a well-defined definition of critical infrastructure in place. According to "Public Safety Canada" the national critical infrastructure is defined as "processes, systems, facilities, technologies, networks, assets and services essential to the health, safety, security or economic well-being of Canadians and the effective functioning of the Canadian government" (Public Safety Canada, 2019a). The critical infrastructure has been identified across the 10 key sectors in Canada:

The Canadian government since 2010 has launched a set of initiatives including *National Strategy for Critical Infrastructure (The Strategy)* and *Action Plan for Critical Infrastructure (The Blueprint for Implementation).* Furthermore, the responsibilities for the protection and resilience of the critical infrastructures are shared by the federal, provincial and territorial governments, local authorities and critical infrastructure owners, and operators. In addition, in case of a disruption, individual Canadians are also expected to be prepared for at least the first 72 hours of an emergency.

Sector	*Sector-specific federal department/agency*
Energy and utilities	Natural Resources Canada
Information and communication technology	Innovation, Science and Economic Development Canada
Finance	Finance Canada
Health	Public Health Agency of Canada
Food	Agriculture and Agri-Food Canada
Water	Environment and Climate Change Canada
Transportation	Transport Canada
Safety	Public Safety Canada
Government	Public Safety Canada
Manufacturing	Innovation, Science and Economic Development Canada; Department of National Defence

Figure 18.1 Critical infrastructure sectors and lead federal departments/agencies

(Source: Action Plan for Critical Infrastructure (Public Safety Canada, 2010b))

National Strategy for Critical Infrastructure (The Strategy)

The National Strategy for Critical Infrastructure was first published in 2010 and continues to guide and enhance the overall approach to CIPR in Canada. The strategy is approved by the Federal/Provincial/Territorial (FPT) Ministers responsible for EM in their regions. The strategy has three fundamental concepts and guiding principles (collaboration, all-hazards approach, and information sharing) (Public Safety Canada, 2010a) and is directly guided by the Canadian Emergency Management Framework.

The Emergency Management Framework is complementary to each local government's jurisdiction. Each federal, provincial, and territorial government has its own core responsibility for emergency management and public safety in their jurisdictions (Public Safety Canada, 2017). Furthermore, once every five years, the framework is subject to review and revision by the federal, provincial, and territorial governments to ensure that it evolves with the constant change and remains accurate and relevant. The framework outlines a collaborative approach for federal, provincial and territorial emergency management initiatives and it is comprised of interdependent risk-based functions: "prevention, mitigation, preparedness, response and recovery" (Public Safety Canada, 2010a).

Action Plan for Critical Infrastructure (The Blueprint for Implementation)

To effectively address the evolving risk environments and the implementation of the *National Strategy for Critical Infrastructure*, the Action Plan for Critical Infrastructure has three strategic objectives:

- Building partnerships
- Sharing and protecting information
- Implementing an all-hazards risk management approach (Public Safety Canada, 2018a).

Under each of the three strategic objectives, the action plan sets out tangible initiatives focused on promoting a collaborative approach among federal, provincial and territorial governments, and critical infrastructure owners and operators to identify, manage, and mitigate risks to critical infrastructures before they lead to disruptions. Furthermore, the action plan defines key areas of collaboration for the federal, provincial and territorial governments, and critical infrastructure owners and operators.

As part of it is a core strategy, Public Safety Canada has developed the CIMF to address some of the cross-provincial-boundary elements of cyber security, The CIMF was designed to complement the existing federal, provincial, and territorial emergency management frameworks and plans as well as emergency management plans from the critical infrastructure owners and operators.

In summary, the National Strategy and the *Action Plan for Critical Infrastructure and Cyber Incident Management Framework* (CIMF) together establish a strong risk-based baseline to advance coherent and coordinated actions of federal, provincial, and territorial initiatives – among the ten critical infrastructure sectors for strengthening the resiliency of Canada's critical assets and systems such as health, agriculture, power grids, transportation, information and communication networks, and public safety systems.

18.2. Implementation

The core of the Canadian Emergency Management Framework for the CIP is built around three key guiding principles: building partnerships, sharing and protecting information, and implementing an all-hazards risk management approach. The Canadian mandate for critical infrastructure protection comes from the provisions of the Emergency Management Act, 2007, the *National Strategy for Critical Infrastructure and Canada's Cyber Security Strategy* (Government of Canada, 2016). The Emergency Management Framework was last updated in 2017 (Public Safety Canada, 2019b). Federal, provincial, and territorial (FPT) governments operate cooperatively within the jointly established FPT emergency management governance structure, which facilitates coordination and collaboration in full respect of each government's legislated jurisdiction (Public Safety Canada, 2017).

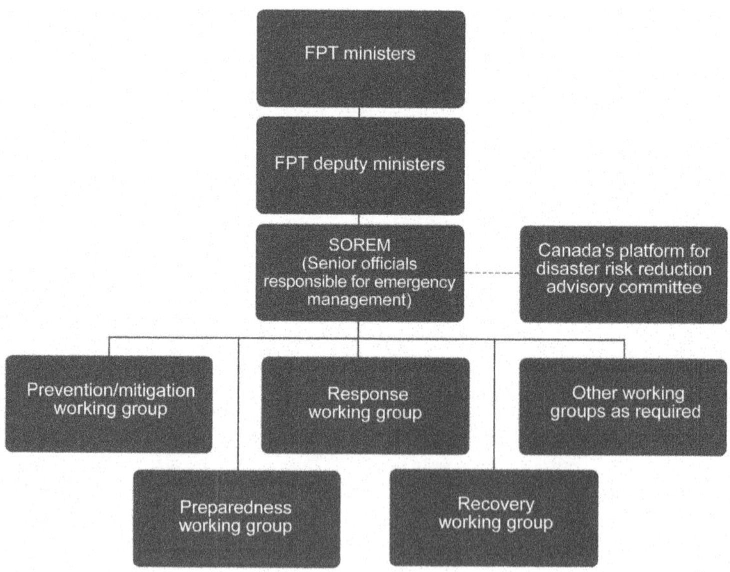

Figure 18.2 Federal, provincial, and territorial (FPT) governance structure

(Source: An Emergency Management Framework for Canada (Public Safety Canada, 2017))

Building Partnerships

The multitude of emergencies in Canada is mostly managed by the municipalities and communities, or at the provincial or territorial level (Government of Canada, 2011). Therefore, each federal, provincial, and territorial government has its own core responsibility for the EM and the public safety in their jurisdictions (Public Safety Canada, 2017). Public Safety Canada, together with Lead Federal Departments

Actor	*Role*	*Responsibilities*
Federal government	Lead federal activities	• Advance a collaborative federal, provincial, and territorial approach to strengthening the resiliency of critical infrastructure • Collaborate with provincial and territorial governments to achieve the objectives of the strategy • Collaborate with national associations • Collaborate with critical infrastructure owners • Operators within federal mandate in consultation with provinces and territories
Provincial/ territorial governments	Lead provincial/ territorial activities	• Advance a collaborative federal, provincial, and territorial approach to strengthening the resiliency of critical infrastructure • Collaborate with federal, provincial, and territorial governments to achieve the objectives of the strategy • Coordinate activities with their stakeholders, including municipalities or local governments, where it applies, associations and critical infrastructure owners and operators
Critical infrastructure owners/ operators	Collaboratively manage risks related to their critical infrastructure	• Manage risks to their own critical infrastructure • Participate in critical infrastructure • Identification, assessment, prevention, mitigation, preparedness, response, and recovery activities

Figure 18.3 Roles and responsibilities

Source: Action Plan for Critical Infrastructure (Public Safety Canada, 2010b)

(LFD), is working narrowly with federal, provincial and territorial governments, local authorities, and critical infrastructure owners to establish sector networks.

To further enhance collaboration across Sector Networks, the National Cross Sector Forum (NCSF) was established. The NCSF is a national-level consultation and outreach entity that brings together leaders from Canada's 10 critical infrastructure sectors to identify priorities and discuss cross-sector issues and initiatives to enhance the resilience of Canada's vital assets and systems. In addition, Public Safety Canada is closely working together with its counterpart The Homeland Security in the USA to develop and implement policies and procedures to protect its critical infrastructures. The policies and procedures are heavily inspired by the NIST framework used in the USA (Public Safety Canada, 2016).

Moreover, the Regional Resilience Assessment Program (RRAP) across Canada was established. The program includes the development of a cyber security assessment tool and facilitates multisector network meetings and exercises to identify interdependencies and address cross-sector issues. The RRAP uses mainly three assessments tools, The Critical Infrastructure Resilience Tool (CIRT), the Critical Infrastructure Multimedia Tool (CIMT), and the Canadian Cyber Resiliency Review (CCRR) to help organizations measure and improve their resilience to all hazards across Canada (Public Safety Canada, 2018b).

To further enhance international partnerships, the Emergency Management Framework is also in alignment with key international agreements such as Sendai Framework for Disaster Risk Reduction 2015–2030 to further enhance disaster risk reduction priorities globally. The Sendai Framework is a 15-year, voluntary, non-binding agreement that recognizes that the State has the primary role to reduce disaster risk but that responsibility should be shared with other stakeholders including local government, the private sector, and other stakeholders. The framework also supports the implementation of the Paris Agreement, aimed at reducing the impact of climate change, and the United Nations Sustainable Development Goals (SDGs) to help Canada enhance its approach to EM and disaster risk reduction. Public Safety Canada will also continue to participate in a number of international groups, such as the Critical Five (Canada, Australia, Great Britain, New Zealand, and United States of America). These groups have been formed to provide a forum for the discussion of critical infrastructure resilience issues of mutual interests.

Sharing and Protecting Information

In order to strengthen critical infrastructure resilience, effective sharing and protection of information on a "need to know principle" is required. Furthermore, timely sharing of critical information across governments and critical infrastructure sector stakeholders is paramount to effective risk management. The national *Counter-Terrorism Strategy* stresses the importance of cooperation and sharing of information with Canada's domestic and international partners at all levels of government, intelligence and law enforcement agencies, industry stakeholders, and civil society (Public Safety Canada, 2018c).

To establish and maintain governance structures and mechanisms to facilitate effective cooperation, Public Safety Canada continues to work closely with critical infrastructure stakeholders, governments, and security and intelligence organizations, e.g. Royal Canadian Mounted Police (RCMP), Canadian Security Intelligence Service (CSIS), Canadian Border Services Agency (CBSA), and the Canadian Cyber Incident Response Centre (CCIRC) to assess the evolving threat environment and provide relevant information to the critical infrastructure community (Public Safety Canada, 2016).

Furthermore, Public Safety Canada facilitates effective information sharing for critical infrastructure stakeholders, governments, and security and intelligence organizations through Canadian Cyber Threat Exchange (CCTX)[1], Virtual Risk Analysis Cell (VRAC)[2], CCIRC[3], and Canadian Critical Infrastructure Information Gateway. The CI Gateway is a forum for CI owners and operators for timely information sharing and secure collaboration across 10 key critical infrastructure sectors. Members can securely access documents, links, RSS feeds, geospatial material, and contact information from every CI sector quickly and efficiently. Users can also recommend and submit relevant content to be shared with the CI Gateway community (Public Safety Canada, 2016).

18.3. Public-Private Partnership

As with much critical infrastructure throughout the world, more than 85% of Canada's infrastructure is owned and operated by the private sector (Government of Canada, 2004). Therefore, strengthening the resilience of critical infrastructure requires public-private partnerships among all partners and stakeholders. Public Safety Canada works closely with federal departments and agencies, provinces and territories, private sectors, and international counterparts to build partnerships. Initiatives such as RRAP, Cross-Sector Exercises, Sector Networks, and NCSF are focused on building, sustaining, and enhancing collaboration with all partners within the critical infrastructure community. Moreover, Public Safety Canada also recently launched the Cyber Security Cooperation Program (CSCP) as a means to improve the security of Canada's vital cyber systems (Public Safety Canada, 2016).

Furthermore, Public Safety Canada has invested to promote a coordinated, global approach to critical infrastructure. As an example, Canada and the United States of America have achieved great results toward meeting shared objectives of enhancing the resilience of critical infrastructure through the *Canada-United States Action Plan for Critical Infrastructure* (Canada–U.S. Action Plan) (Public Safety

Canada and U.S. Department of Homeland Security, 2010). The core purpose of the Canada–U.S. Action Plan is to strengthen the safety, security, and resiliency of Canada and the United States by establishing a comprehensive cross-border approach to CIPR and working across sectors and jurisdictions. Moreover, the partnership will allow Canada and the United States to effectively and efficiently address a range of cross-border critical infrastructure issues and enhance collaboration to share best practices, identify interdependencies, and conduct joint exercises to further improve safety, security, and resiliency of the critical infrastructure in both nations.

Canada has also closely worked together with the broader international community, including Five Eyes partners (US, UK, Australia, New Zealand), to ensure that its approach to strengthening CI resilience reflects the global context and builds on the best practices from trusted allies. The Resiliency Council of Canada together with its Transatlantic partners is conducting and disseminating advanced research in the domains of security and resiliency of CI on Land, Sea, Air, Space and Cyberspace.

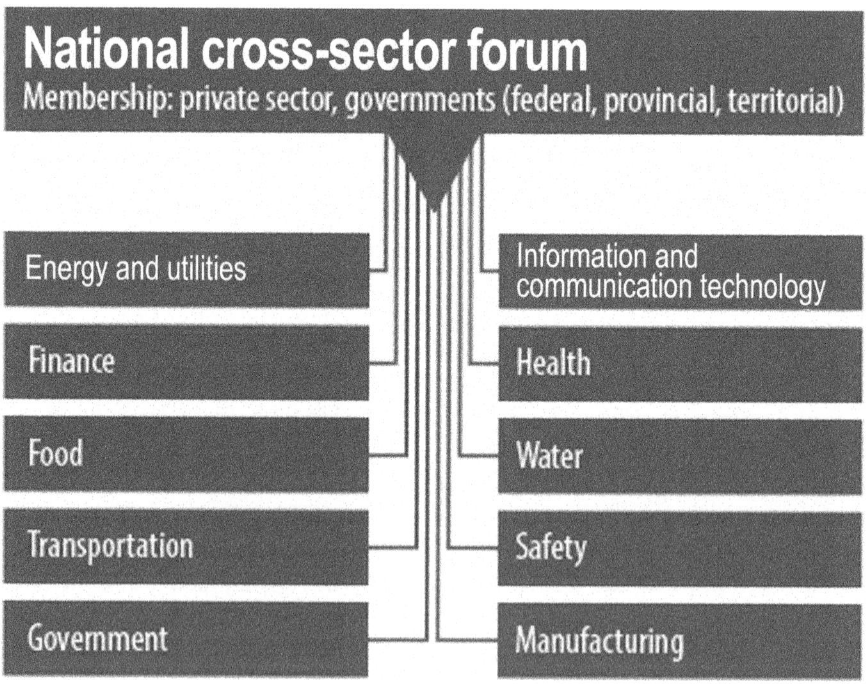

Figure 18.4 Sector networks and the national cross-sector forum

(Source: National Strategy for Critical Infrastructure (Public Safety Canada, 2010a))

18.4. Cooperation with European Commission

To effectively and efficiently protect the CI in Canada from emerging threats, the Government of Canada together with its Transatlantic partners such as North Atlantic Treaty Organization (NATO) and the European Union (EU) is developing advanced cyber capabilities. Canada has been a third-country contributor to EU CSDP civilian and military missions since 2003. The EU and Canada remain close, like-minded partners in the global fight against terrorism. Both the EU and Canada contribute to the fight against terrorism in the framework of the Global Coalition against Daesh. The last annual EU-Canada Dialogue on Counterterrorism was held on 6 July 2017 in Ottawa to exchange views on counterterrorism and on preventing and countering violent extremism (CVE). As part of the external dimension of the EPCIP, the EU and Canada exchange information and good practices (Government of Canada, 2018).

Furthermore, the EU's Permanent Structured Cooperation (PESCO) initiative has put cyber security at the top of the priority list for enhanced collaboration and cooperation among Member States, neighboring countries, and it is Transatlantic partners the US and Canada. On March 20, 2019, at the 5th European Union Security and Defence Symposium held in Ottawa, Canadian and EU government officials among federal, provincial and territorial governments, local authorities, and CI owners and operators shared the progress being made on countering cyber threats via CRRT.

The CRRTs will allow the Member States, EU Institutions as well as Transatlantic partners to help each other to ensure a higher level of cyber resilience and coordinated response to emerging cyber threats. CRRTs will be equipped with commonly developed deployable cyber toolkits designed to detect, recognize, and mitigate cyber threats. Moreover, the CRRT project will further enhance Canada and the EU cooperation and strengthen the shared responsibilities in the field of emerging cyber threats (PESCO Secretariat, 2019).

18.5. Education and Training

A strong security awareness program is key to keeping up with the ever-changing cyber security battleground (Public Safety Canada, 2016). To address risks associated with security awareness related to the convergence of physical and cyber critical infrastructure systems, Public Safety Canada is providing training sessions on how to protect ICS and is bringing together stakeholders to share their knowledge and experience on mitigating cyber threats related to CI. The ICS security workshops bring together recognized experts along with representatives from the federal, provincial and territorial governments, local authorities, and CI owners to provide briefings on the latest threats and steps that can be taken to increase the security of ICS in their environments. ICS security training is focused on the development of basic incident handling and response skills for the ICS environment. The training is hands-on using real tools and targets.

To further raise security awareness at all levels in organizations across Canada and to enhance understanding of cyber-risks related to CI, TRCC is offering more than 22 certified tailored cybersecurity awareness courses, training courses, and board briefings. Participants can discuss Cyber Security issues, share knowledge, and prioritize actions to address threats, risks, and vulnerabilities to address CI security challenges.

Moreover, CI owners and/or operators who need assistance in assessing their organization's security awareness have access to the RRAP. The RRAP provides the CCRR to help organizations measure and improve their resilience to all hazards across Canada (Public Safety Canada, 2018b). The CCRR is a one-day, on-site facilitated workshop designed to assess an organization's overall approach to cyber security practices and procedures.

Chapter Conclusion

Normative Framework

To shape the national CIP frameworks and to reflect the emerging nature of EM, the Government of Canada has the National Strategy, the Action Plan for CI and CIMF in place. The strategy, action plan and CIMF, together, establish a strong risk-based baseline to advance coherent and coordinated actions of federal, provincial, and territorial initiatives and among the 10 CI sectors for strengthening the resiliency of Canada's critical assets and systems. Moreover, they also guide the development of a comprehensive and flexible normative framework to effectively and efficiently identify systems, networks, and objects of national importance and implement mitigation strategies. In addition, they provide a risk-based approach to understanding the importance of CI service stability and continuity, CIP, cyber security, information sharing between federal departments, collaboration with international and private sector partners, and protection of sensitive private sector information.

Public-Private Partnerships

Canadian Federal, Provincial and Territorial (FPT) governments operate cooperatively within the jointly established FPT emergency management governance structure, which facilitates coordination and collaboration in full respect of each government's legislated jurisdiction. Public Safety Canada together with LFD is working narrowly with federal, provincial and territorial governments, local authorities, and CI owners to establish Sector Networks.

To further enhance collaboration across Sector Networks, the NCSF was established. The NCSF is a national-level consultation and outreach entity that brings together leaders from Canada's 10 CI sectors to identify priorities and discuss cross-sector issues and initiatives to enhance the resilience of Canada's vital assets and systems. In addition, Public Safety Canada is closely working together with its counterpart the Homeland Security in the USA to develop and implement policies and procedures to protect its CIs.

Sharing and Protecting Information

The RRAP across Canada includes the development of a cyber security assessment tool and facilitates multisector network meetings and exercises to identify interdependencies and address cross-sector issues. The RRAP uses mainly three assessments tools, The Critical Infrastructure Resilience Tool (CIRT), the CIMT, and CCRR to help organizations measure and improve their resilience to all hazards across Canada. Public Safety Canada continues to work closely with CI stakeholders, governments, and security and intelligence organizations e.g. RCMP, CSIS, CBSA, and CCIRC to assess the evolving threat environment and provide relevant information to the CI community Public Safety Canada facilitates effective information sharing for CI stakeholders, governments, and security and intelligence organizations through CCTX, VRAC, CCIRC, and Canadian Critical Infrastructure Information Gateway.

International Collaboration

Canada has also closely worked together with the broader international community, including Five Eyes partners (US, UK, Australia, New Zealand), to ensure that its approach to strengthening CI resilience reflects the global context and builds on the best practices from trusted allies. The Resiliency Council of Canada together with its Transatlantic partners in Europe is conducting and disseminating advanced research in the domains of security and resiliency of CI on land, sea, air, space, and cyberspace. Furthermore, the Government of Canada together with it is Transatlantic partners such as NATO and with the European Union (EU) is developing advance cyber capabilities. Canada has been a third-country contributor to EU CSDP civilian and military missions since 2003. Moreover, the EU's PESCO initiative has put cyber security at the top of the priority list for enhanced collaboration and cooperation among member states, neighboring countries, and its Transatlantic partners USA and Canada.

Security Education Training and Awareness

To address risks associated with security awareness related to the convergence of physical and cyber CI systems, Public Safety Canada is providing training sessions on how to protect ICS and is bringing together stakeholders to share their knowledge and experience on mitigating cyber threats related to CI. The ICS security workshops bring together recognized experts along with representatives from the federal, provincial, and territorial governments, local authorities, and CI owners to provide briefings on the latest threats and steps that can be taken to increase the security of ICS in their environments. ICS security training is focused on the development of basic incident handling and response skills for the ICS environment. The training is hands-on using real tools and targets.

To further raise security awareness at all levels in organizations across Canada and to enhance understanding of cyber-risks related to CI, TRCC is offering more than 22 certified tailored cybersecurity awareness courses, training and board briefings. Participants can discuss cyber security issues, share knowledge, and prioritize actions to address threats, risks, and vulnerabilities to address CI security challenges.

Notes

1. CCTX is an independent, not-for-profit organization, whose mission is to share threat information, conduct cyber threat analysis and recommend risk mitigation measures.
2. Provides expertise and analysis to identify the potential impacts of disruptive events and support improved planning and swift response and recovery when incidents occur.
3. Provides a range of guidance documents, security bulletins, and technical reports related to cyber security issues as well as timely and actionable information in response to cyber events.

5 OVERVIEW AND CONCLUSION

DOI: 10.4324/9781003273769-23

Chapter 19

Analysis of All Countries' Approaches

Alessandro Lazari, Robert Mikac, and Ivana Cesarec

Contents

This chapter provides an insight into the overall analysis of all country approaches case studies. This chapter aims to draw key observations for each of the units within each chapter and extract all the essentials that the authors have emphasized in the analyses they have made. At the same time, the goal is to show what policies and processes have been implemented and what has yet to be done. Each of the analyzed units within this chapter begins concerning the state of affairs in the European Union. Because the Union was chosen as the initial indicator of the research, all chapters were written with the development and achievements of strengthening the resilience and protection of CI concerning the Union. This is particularly pronounced in countries joining full membership because, before the official accession to the Union, it is required to implement the EU *Acquis Communautaire* (which includes policies for the protection of CI) in national legislation.

DOI: 10.4324/9781003273769-24

19.1. Introduction

The beginning of the analysis of a certain area is best started by reviewing the strategic documents that a certain organization or state has in that area. Based on the Founding Treaties and their Amendments, which represent its primary legislation, the European Union has developed the following set of strategic documents that provide clear visions and guidelines for developing all key areas. Security strategies are most relevant to our consideration. Listed are *European Security Strategy: A Secure Europe in a Better World* (2003), *Internal Security Strategy for the European Union: Towards a European Security Model* (2010), *Cybersecurity Strategy of the European Union: An Open, Safe and Secure Cyberspace* (2013), *The Global Strategy for the Foreign and Security Policy of the European Union (2016*), and *EU Security Union Strategy* (2020). In the strategies, the area of CI has gained more and more attention and space over time, and ultimately in the last strategy where CI was recognized as an indispensable component of the functioning and development of the Union, Member States, economy, and European way of life.

Furthermore, it is vital to draw the focus of research. That is the EU Critical Infrastructure Protection External Dimension area, which is part of the EU Enlargement Agenda, EU EaP, and Transatlantic cooperation. Critical Infrastructure Protection External Dimension is one of the fundamental pillars of the development of this area from the Union aspect. The EU Enlargement Agenda is a set of individual policies toward seven states (Turkey, BiH, Serbia, Montenegro, North Macedonia, Albania, and Kosova) in their process of accession to the Union. This cooperation framework was established to formalize the Union's cooperation and assistance to the seven countries. The EU EaP is a joint policy initiative between the European Union, its Member States, and its six Eastern neighbors: Ukraine, Belarus, Moldova, Georgia, Armenia, and Azerbaijan. The purpose of this partnership is in the areas of trade, mobility, economic development, and better quality of life. While Transatlantic cooperation includes a partnership, consultations, and cooperation with two transatlantic partners (the United States, and Canada). The cooperation covers several areas of interest. There are certain common denominators of collaboration in all three cooperation packages between the Union and the groups of countries mentioned here. One of them is the area CI. In addition to analyzing the state of regulation of CIP areas in each country, this research's central interest is on cooperation between the Union and each country.

The next step was setting the research framework because we presented Croatia as a role model for considering the development of national legislation, procedures and activities, and cooperation with the Union before achieving full membership and after. In this way, we wanted to show how Croatia regulated its CIP area and cooperation with the Union, which served as a guide for all our contributors to analyze their countries. This was much more applicable to countries within the EU Enlargement Agenda and the EU EaP, to a lesser extent to Transatlantic cooperation. Before full membership in the Union (full member since 2013), Croatia recognized the concept of CI in several strategic documents (*National Strategy for the Prevention and Countering of Terrorism* (2008), *Protection and Rescue Plan for the Republic of Croatia* (2010), *Risk Assessment for the Republic of Croatia from Natural and Technical-Technological Disasters and Major Accidents* (2013), *National Strategy and Action Plan for the Non-proliferation of Weapons of Mass Destruction* (2013)). The initial definition of CI was given in the *Private Protection Act* (2010). At the same time, the entire area was regulated by laws and bylaws in the field of internal security and defense. We can say that Croatia had a solid strategic and normative basis before the introduction of a new concept based on the European Program for Critical Infrastructure Protection policy package (with *Council Directive 2008/114/EC* as the most significant document). A new and comprehensive approach to CIP began with a review of the obligations under *Council Directive 2008/114/EC*, which resulted in the drafting and adoption of the *Critical Infrastructure Law* in 2013, just before full membership. Under the said Law, rules and regulations were drafted and adopted in the same year, which enabled the implementation of the Law's provisions in practice. Regarding, EU Cybersecurity policy package (with the most relevant document, *Directive (EU) 2016/1148* (NIS Directive)), Croatia has implemented all the required provisions as well. Both segments will be explained in more detail in the next subchapter.

Considering the countries included in the EU Enlargement Agenda, the strategic and normative framework positions that relate directly or indirectly to the CIP area are quite different. This clearly defines the further possibilities and activities that each country can implement in a short- and medium-term plan related to protecting CI through a systemic approach.

Turkey has implemented several CIP strategic efforts, particularly to strengthen the CI cybersecurity, such as implementing *National Cyber Security Strategies* (2013–2014 and 2016–2019), the national Cyber Incidents Response Team, and mandates to establish Sectoral Cyber Incident Response Teams. The focus of consideration is on the cyber domain, which is not uncommon. Similar examples exist in some highly developed EU countries (such as France), where the CIP's primary focus is in the cyber domain. The next important thing to point out is that Turkey focuses on protecting CI within energy-related networks. It is also important to note that despite the increased interest and accelerated efforts to secure cyberspace in Turkey, there is currently no legislation that "directly" and specifically addresses the CIP.

As far as BiH is concerned, before any deeper insight, it is necessary to emphasize the state's territorial-political specificity and its way of functioning, which is unique in comparison to all other analyzed countries. In BiH and the state itself, three regional territorial-political entities have their own constitutions, governments, assemblies, and enacted laws and bylaws. Then, at the level of one of the regions within its 10 subregional cantons, there are governments and assemblies that pass laws and bylaws. It is important to point out that such a state structure opens numerous challenges in all areas, including the CIP. Ultimately, there are a few areas where a synergistic effect of different political levels has been observed in the creation of the CIP framework. There are no relevant strategies or normative acts at the state level that could be pointed out as directly related to the CIP area. While the closest to our area of interest are assessments, plans, and laws related to disaster risk reduction and civil protection. However, it is interesting to point out that one of the entities, the Republic Srpska, passed the *Critical Infrastructure Law* in 2019, which is the beginning of the CIP system's construction in its area. At the state level, there are still no initiatives or plans for such an Law.

Serbia has a very similar approach and solutions to Croatia. The CIP concept was presented in Serbia for the first time in 2011, as a part of the *Regulation on the Content and Methodology for the Development of Protection and Rescue Plans in Emergency Situations*. Since then, it has been gradually introduced into key security strategies (*National Strategy of Protection and Rescue in Emergency Situations* (2011) and *National Security Strategy* (2019)), in normative documents (*Data Secrecy Law* (2009), *Law on Planning and Construction* (2009), *Law on Water* (2012), *Law on information security* (2016), *Law on disaster risk reduction and emergency management* (2018), *Law on Private Security* (2018), *Law on Environmental Protection* (2018)), as well as in the laws in the field of internal security and defense. All of the above were sectoral approaches and solutions. Serbia adopted the *Critical Infrastructure Law* in 2018, which marked the beginning of a systematic approach to this topic and area.

Montenegro is on a very similar track, linking the initial steps of developing the CIP concept to the area of disaster risk reduction and civil protection and introducing it gradually into strategies and laws. Of all analyzed countries, Montenegro has the largest opus of strategies (*Strategy for disaster risk reduction* (2018), *National Security Strategy* (2018), *Strategy on cybersecurity* (2018)) which have direct guidelines for the development of the CIP area. Also, an enviable number of sectoral laws in which CI is represented (*Law on protection and rescue* (2007), *Law on state property* (2009), *Law on cybersecurity* (2016), *Law on protection of persons and objects* (2018)). As in previous cases, the various sectoral policies for protecting CI have been consolidated through the systemic normative document for the CIP area – *Critical Infrastructure Law*, which was adopted at the end of 2019.

As for North Macedonia, its situation is such that significant strategic and normative activities have been undertaken to emphasize the importance of CI and the need for adequate protection. The concept of CI (not necessarily under that name and definition) is stated in the *National Cyber Security Strategy* (2018), as well as in the following laws: *draft Defense Law*, *Law on Internal Affairs*, *Crisis Management Law*, *Protection and Rescue Law* and *Law on Private Security*. For the efforts made so far in North Macedonia regarding the regulation of CI in general, and the protection of CI, in particular, it is important to point out two things. First, North-Macedonian experts believe there is a lack of clear specification of the term "critical infrastructure". Which clearly complicates every further step. Secondly, it is important to note that activities in creating the framework for defining, legal solutions and implementation of the CIP area are still not in line with the guidelines (and future obligations for all full Member States and those wishing to become one) under *Council Directive 2008/114/EC.*

Albania has so far regulated its strategic determination toward the CIP area within three related frameworks. The first refers to the *National Security Strategy* (2014), which highlighted the importance of CI in general and the need to adopt a national strategy for cybersecurity to provide clear guidelines toward the cyber field. The second framework refers to the integration processes that Albania is implementing at the regional and European levels. It is related to Albania's participation in the EnC and the TeC. Both communities are international organizations with legally binding frameworks that bring together the European Union and its neighbors to create an integrated pan-European energy market related to the first community and extend the EU transport market rules, principles, and policies related to the second community. Although all the countries analyzed so far are full members of these communities (except Turkey which is the Observer in the first community, while it does not participate in the second), Albania is the only one that sees a direct benefit for the CIP area. The third important framework is the gradual adaptation of Albanian legislation to solutions coming from the EU level. Thus, part of the provisions of *Council Directive 2008/114/EC* is integrated into *Law on civil protection* (2019), as well as part of the provisions of *Directive (EU) 2016/1148* (NIS Directive) implemented in *Law on cybersecurity* (2017).

Regarding Kosova, it should be said that it is the youngest country in Europe that gained its independence in 2008 and is still a significant international assistance recipient. It has various UN, NATO, EU, and OSCE missions on its territory, which participate, among other things, in building Kosova's institutions. Kosova still faces significant challenges in recognition of its international status. So far, it has not been recognized by half of the total UN Member States, as well as some of the EU countries (such as Spain). This is certainly important for reviewing the development of the CIP area in Kosovo. Even in such circumstances, Kosova's institutions have achieved an enviable level of development and adoption of strategic and normative documents, which, in addition to other central concern areas, have a link with the CIP. When we consider strategies, we can refer to three *Counter-Terrorism Strategy and Action Plans* (2009, 2012, 2018) and the *National Cybersecurity Strategy and Action Plan* (2016). While the following laws are significant: *Law on Preventing and Fighting Cybercrime* (2018) and *Law on cybersecurity* (2020). When we consider the specifics of the CIP area, it needs to be pointed out that Kosova has transposed the *Council Directive 2008/114/EC* into national legislation by adopting the *Law on critical infrastructure* in 2018 which entered into force in April 2019.

Analyzing these countries together, we have a situation where they aspire to become a full member of the European Union. Four of them are full members of NATO (Turkey since 1952, Albania since 2009, Montenegro since 2017, and North

Macedonia since 2020), which is certainly an added value when regulating the CIP area. The two states (BiH and Kosova) have certain significant territorial and political specifics of the way other states are organized and function. To a greater or lesser extent, all countries have adopted strategic documents in the field of security in which they have recognized the importance of the CIP area. In these strategic documents, the CIP concept's general framework is considered in most countries, while in Turkey, it is primarily directed toward the cyber domain. Regarding normative documents, three countries have adopted a systemic law on CI (Serbia, Montenegro, and Kosova), in BiH there is such a law but at the regional level. Albania has partially regulated this area through other laws. In contrast, Turkey and North Macedonia do not have a legislative framework in place that directly and specifically addresses CIP.

We will start the analysis of countries within the EU EaP initiative with Ukraine. Compared to all the countries analyzed so far, Ukraine has particular challenges and security threats related to terrorist threats, sophisticated cyber-attacks against infrastructure, and hybrid aggression. On the one hand, this makes it difficult for her to function, while on the other hand, it forces her to be more agile in protecting CI. Although Ukraine has no formal obligation to regulate the CIP area based on EU guidelines, directives, and policies, it has become an orientation for Ukraine's internal efforts in developing a legislative basis for a CIP policy. The parts of the *National Security Strategy* (2015) and *Cybersecurity Strategy* (2016) related to the CIP area have been developed in full compliance with the EU *Council Directive 2008/114/EC* and *Directive (EU) 2016/1148* (NIS Directive). Based on these documents as well as the strategic orientation of building a CIP system inspired by and is correlated with approaches reflected in EU legislation, Ukraine adopted the *Law on cybersecurity* (2017), *Concept for building a state critical infrastructure protection system in Ukraine (2017)* and drafted *Law on critical infrastructure and its protection* which is currently in the harmonization phase.

In the introductory part of Belarus, it is necessary to point out three important things for considering the CIP area. First, Belarus has several strategic and normative documents that mention CI and critical objects or documents linked to CIP activities in a broader context. These are the following strategic documents: *National Security Concept* (2010), *Concept of Energy Security* (2015), *Strategy of Informatization Development in Belarus* (2015), *Information Security Concept of the Republic of Belarus* (2019). Also, several laws mentioned CI but did not deal with it in essence. This brings us to the second important thing in looking at the Belarus CIP area. Belarus has not implemented any specific, consolidated CIP law, nor is there currently a plan to enact such law. Also, Belarus lacks a strategic document that would indicate or determine the development of this area. The third important thing is that although Belarus currently does not have the plan to draft a systemic *Law on critical infrastructure*, this does not mean that it is not being considered. Belarus seeks legislative inspiration for this in EU documents and best practices and in activities implemented by the CIS organization and the CSTO (Belarus is a member of both organizations).

Moldova has a very similar situation as Belarus in terms of strategic and normative documents concerning CIP activities' regulation. The documents exist, touch on, and mention CI, but none of these documents is systematically and specifically focused on CIP. In Moldova, it is necessary to note the geopolitical challenges in the adoption of strategic documents. This is a situation that none of the analyzed countries has had so far in this analysis. Moldova developed during 2015–2016 the *National Security Strategy* (adopted in 2016), with NATO experts' involvement. The strategy recognized CI as an area and a concept, but the strategy underlined the links with NATO and indirectly viewed Russia as a security risk source. New president Igor Dodon (2016–2020), with pro-Russian sympathies, withdrew the strategy via the repealing procedure one year later. So Moldova does not currently have a valid *National Security Strategy.* Other strategic documents that are relevant to consider and address to some extent to the CIP are the *National Cyber Security Program* (2016), *Concept of Information Security* (2017), *National Defense Strategy and Action Plan* (2018), *Informational Security Strategy* (2019). At the legislative, law level, the legal framing of the protection of CI came with the *Law on prevention and combating terrorism* (2017). What is currently most important is to draft the *Law on the protection of essential infrastructure.*

We begin our analysis of Georgia by stating that there is no official definition of CI (but there is a term critical information infrastructure) in Georgian laws. Strategic documents are not included in this analysis, assuming that they are not linked to CI. Further, CI and CIP are not regulated by legislation. As Georgia has no formally defined list of CIs, there is no institutional division of responsibilities for CIP. Nevertheless, like any other country in the world, Georgia has objects, processes, and systems strategically important to its functioning. It protects them through various laws but is called by some other name, not CI. This is identical to the example of North Macedonia that we analyzed earlier. What is void in the European Union and its Member States in defining CI, in Georgia is mentioned within various laws under terms such as objects of strategic and special importance and the property of special importance. Among the most important laws regulating Georgia's "critical infrastructure" are the *Law of State Secrets* (1996), *Law on Privatization of State Property, Privatization of Property of Local Self-Government Unit and Transfer of Right to Use* (1997), *Law on Engineering-Geodesic Control and Objects of Strategic and Special Importance* (2003), *Law on State-Owned Property* (2010), and *Law on Information Security* (2012). None of these laws are systemic and comprehensive that they regulate the entire area of CIP.

Armenia has a very similar situation to Georgia, related to terminology, that is, definition of the term and concept of critical infrastructure. There are objects and essential systems, but it does not call them "critical infrastructure". Armenia is a country whose sources of insecurity dominate a conventional military character (the last such case was the very recent short-lived war between Armenia and Azerbaijan in autumn 2020). So its observation of the concept of CI is conditioned by this reality. This was expressed in the *National Security Strategy* from 2017, as in the current *National Security Strategy* from 2020. In the new strategy, CIP is placed in the context of deterrence and neutralization of military threats and ensuring open and safe information and cyber domains. Within the laws, special objects and objects of strategic importance are often used as a synonym for CI. The protection of special and objects of special importance is carried out according to national legislation and international conventions.

Azerbaijan has still a long roadmap ahead since CIP is not yet formally defined in the national framework even though the country faces serious challenges to the national CI which are at the same time vital for the entire area which covers from the Caspian region up to the Mediterranean one. The closest initiatives in the area of CIP have focused on energy, which is rather strategic, and in the area of cybersecurity.

The CIP analysis of the US and Canada has a completely different dimension and character than the previous countries. Therefore, this whole section is written differently than the previous ones. The strategic and normative deficit has been expressed to a greater or lesser extent in all analyzed countries concerning essential and systemic CIP solutions. For the USA and Canada, we can initially state that this is not the case. These are the countries that have a very well regulated CIP area and one of the few countries (in this category we can include only a few EU members such as Great Britain and France and additionally Australia and New Zealand) that serve as global examples of how is it is necessary to arrange this area. Besides, the United States is a global leader in both the concept of CI and the protection of CI as we know it today. It should also be noted that when the EU began to develop its own approach to the CIP area, legislative inspiration and best practices of US solutions were used, which have already been developed and implemented into national legislation. Therefore, the chapters written by our American and Canadian colleagues, although having the same structure as all analyzed countries, do not list all relevant CIP strategies and laws. The history of development and the current setup of the CIP area are so wide that their analyses should be much longer. For this reason, only indications of the CIP framework within both states and some (but not all) key strategies and laws are mentioned. For this introductory part, we can conclude that the US and Canada have clear definitions of the concept and the term "critical infrastructure", strategic documents that generally guide the CIP area, and specific sector strategies. They have also developed and implemented an appropriate normative framework that prescribes the competencies of acting within the CIP area. The analysis of the specifics, arrangement, and cooperation of different actors will be carried out in the following subchapters.

At the end of the introductory part, it should be noted that the above is written longer than the following subchapters. It was necessary to show the CIP specifics and different development levels of each of the analyzed countries. This is especially important in order to get an introductory picture of the current state of affairs. In the following, the European Union and countries' experiences will be grouped by sections within each subchapter as this book is structured.

19.2. Implementation

This section aims to present the ways and models of implementing CIP solutions in practice. The first section will combine the activities undertaken by the EU and Croatia. The second section will be a joint overview of the activities of the seven EU Enlargement Agenda states. This will be followed by a consolidated analysis of the six EU EaP countries and a review of the US and Canada (Transatlantic cooperation partners).

Based on the visions and strategic guidelines set out in the strategies mentioned in the previous part, the EU normative development and implementation has been initiated and achieved in two parallel and complementary areas: the European Program for Critical Infrastructure Protection policy package (with *Council Directive 2008/114/EC* as the most significant document) and the Cybersecurity policy package (with most relevant, *Directive (EU) 2016/1148* (NIS Directive)). The first policy package has much more substance because it has been developed over a longer period. Nevertheless, both policy packages are very well designed so that the EU institutions and the Member States have clear enough guidelines for implementing the necessary activities in both areas. Likewise, all countries cooperating with the Union in these areas have sufficient material that can serve as a model for them to develop their own approaches to these topics. Regarding the obligations from the European Program for Critical Infrastructure Protection policy package, Croatia adopted the *Critical Infrastructure Law* in 2013, which established a CIP normative framework; has set up the CIP system architecture; and drafted and adopted two bylaws that are important for the process of identification, designation, and protection of CI. Also, various institutions are linked into a single system of CIP; training for security coordinators in the competent central state administration bodies was organized (more about it will be said below); meetings were held with the Republic of Slovenia and Hungary (regarding the identification of ECI) and implementation of the international CIP-focused project (funded by the European Union to support further development of initiated system-building activities). In the next step, Croatia strengthened the entire CIP area by adopting two important national security documents in 2017: the new *National Security Strategy* and the *Homeland Security Law*. In both documents, the protection of CI is set as a national priority. In particular, the strategy underlines the strategic objective of achieving the highest level of security and protection of the population and infrastructure. *The Homeland Security Law* aims to establish an effective Homeland Security System in response to protect the national interests of the Republic of Croatia, taking into account, protection of the people, economy, infrastructure, and natural resources. Related to the EU Cybersecurity policy package *National Cyber Security Strategy* (2015) was adopted concerning the strategic importance of critical communication and information infrastructure and cyber crisis management. The strategy has also provided a general orientation for national cyber priorities. After that, the *NIS Directive* transposed through the *Law on the Cyber Security of the Key Service Operators and Digital Services Providers* (2018). That Law is the upgrade of the existing national cyber architecture, ensured a high level of cybersecurity in providing services, important for carrying out the key social and economic activities. Croatia has made beneficial use of EU solutions from both policy packages to strengthen its own solutions and cooperate as well as possible with other countries and institutions.

For the EU Enlargement Agenda, the situation is as follows. Turkey has a pervasive network of actors involved in the CIP area. Most activities are carried out within the cyber domain, where many institutions operate, drafting implementing acts, guides, and examples of best practices in CIP. As far as the physical protection of CI is concerned, the responsibility lies with the Ministry of the Interior and the Disaster & Emergency Management Authority is responsible for preparedness and crisis management issues in natural and technical disasters. Turkey currently faces several challenges. The first relates to the definition of CI. In Turkey, the term "critical infrastructure" is generally used to address "critical information infrastructure" since most of the national efforts concentrate on strengthening the national cybersecurity framework. There is no supreme authority whose tasks are solely dedicated to CIP, and even there is no national CERT specialized in CI. National CIP is still missing and an all-hazards approach to the whole issue of CIP. BiH also do not have a uniform definition of CI at the state level or an established system for its protection. The specifics of BiH are explained in the previous subchapter. The country is aware of the importance of this area and gradually covers it through certain areas, such as prevention and preparedness, risk assessment of major accidents and disasters, as well as in the field of prevention and suppression of terrorism. Initial preconditions for the development of this area are slowly being created. The term "critical infrastructure" is introduced in various documents, indicating the need for a comprehensive approach to this area and how it is important for BiH. In the absence of a comprehensive approach to this topic, BiH use a sectoral approach, so in some laws of different areas (internal affairs, civil protection) the topic of CIP is integrated – if not explicitly but certainly in essence what CIP means. Serbia is currently in a position to have recently implemented the *Critical Infrastructure Law* and is working on drafting all necessary bylaws. The MoI is the central state institution for developing and implementing the necessary solution. As examples of best practice, it uses models developed within the European Union and its individual Member States. Serbia also draws experience from some CIP international projects. The central focus of the current discussion on CI is about the framework of the future system, the relationship between different levels of government (local and state) and actors within the system, and how best to use the current burst and confidence in building a comprehensive approach to this topic. Regarding Montenegro, they are in a rather similar situation as Serbia. National governance for the CIP system is expected to be developed based on the legal framework of the *Law on determination and protection of critical infrastructure of Montenegro*, which was adopted in December 2019, but implementation has not started yet. The MoI is responsible for the Law's implementation. The next important step is that the Law is complemented by secondary legislation identifying the list of infrastructures considered critical for state security. Intensive work is currently being done on the aforementioned, which will enable the start of the construction of a CIP system in the country. North Macedonia has a rather similar situation as BiH, with the difference that more strategic documents emphasize the importance of the concept of CI and its protection. In the absence of norms that would directly regulate the CIP area (which would mean the beginning of the establishment of a system for CI protection), it is being done in a way that other laws (in the field of defense, internal affairs, private security, civil protection) are adjusted to cover each in its domain CI issues. In North Macedonia, significant activity related to the protection of CI is visible. However, there is a lack of a step that would enact an appropriate law and open this area to establish the CIP system. In Albania, the key impetus for the development of the CI area comes from two reasons. The first concerns the harmonization of national legislation with the provisions of the European Program for Critical Infrastructure Protection policy package and the EU Cybersecurity policy package. In the above, the provisions from both policy packages have been partially transposed into national legislation, although no systemic law on CIP has been drafted so far. The second reason relates to its participation in regional and European energy and transport projects and corridors. So, Albania uses the approach to regulate its internal procedures and activities due to its actions in the international environment. As in all the analyzed countries, Albania is noticeable that it does not have an established single CIP system but performs CIP functions through a sectoral approach. Last but not least, the country we cover in this part is Kosova. Same as Serbia and Montenegro, Kosova has passed the *Law on critical infrastructure protection*. The MoI is the central state institution responsible for the development and implementation of legal provisions. The next similarity is that apart from approving the Law, the normative framework has not yet been completed as secondary legislation has not been drafted, the institutional mechanism has not been established, and the process of identifying and designating CI has not started yet. Nevertheless, what Kosova has, no other analyzed country so far has that. With the TAIEX Expert Mission's support, the MoI developed the *National Critical Infrastructure Plan 2019–2023*, laying down a roadmap that consists of five phases of how the CIP system will be developed in the next few years.

As in the previous group, the six EU EaP countries at different levels of development of the CIP area. At the time of analysis, Ukraine has not yet implemented the unique CIP state management system but is very active in its establishment. The CIP system's main concept was agreed upon, and there is a developed organizational model with a set of principles, mechanisms, and tools to be used. The main subjects/stakeholders within the future CIP system are Government Center on Critical Infrastructure Protection, sectoral authorities responsible for the protection of specific sector CI (10 sectors), functional agencies responsible for security and response to specific threats (six existing security and response systems), local authorities, and operators of CI. Besides, intensive work is being done to develop and implement legal solutions related to protecting physical security and cybersecurity of the CI. Additionally, national critical vital facilities are designed to include facilities identified as the ECI. In developing its own system of CI, Ukraine uses the opportunity to be as close as possible to the line of solutions implemented by the European Union. In Belarus, currently, there is no coordination government body designated to regulate CIP-related issues. In a more general sense, the Belarus Council of Ministers coordinates all the state entities involved in emergency response, emergency preparedness, and protection. At the same time, the main approach to the issues of CIP is of a sectoral nature, where each of the competent institutions takes into account within its sector those facilities, systems, and processes that are critical for the functioning of the state. The highest level of current integration and cross-sectoral approach is in information security, where this area is gradually developing, and key stakeholders are involved. Moldova is in a very similar situation. There are normative and organizational efforts leading to the development of the future CIP system which is currently fragmented and runs in parallel within the following two segments: the non-cyber-related infrastructure and the cyber-related infrastructure. In both areas, there are sector-specific bodies with given authority to develop legal solutions and cooperation between different actors to protect the state's CI. In Georgia, current legislation on national security does not cover any legal bases for CI security, and the existing Georgian legislation does not provide unified regulations on the status of CI. There is a certain "sectoral approach", where sectors define CI objects. Also, there are provisions in Georgian legislation that define CI objects through more "general principles" and their critical status depends on the risk level to national security. The most developed sector within the CIP area is information security, where all required regulations are in place. It should be noted that there is a plan for the recent drafting and adoption of the *Law on Critical Infrastructure Protection* that would encourage additional processes, as well as ensure a unified approach to this topic. Looking at the development of new normative regulations in general and CIP specifically, Georgia directly links the development of its normative framework with the efforts to harmonize the country's legal system with EU legislation, experiences of EU Member States regarding CI, as well as principles and recommendations worked out by international organizations (such as NATO and the United Nations). Like several countries analyzed, Armenia does not currently have a designated central government body with comprehensive authority over developing a normative framework for CIP, the establishment of a CIP system, and the coordination of key institutions and actors. At present, each state ministry implements the actions formulated within the framework of its powers and functions to protect vital facilities, systems, and functions. Armenia is currently in the phase of large-scale reform processes, which have a dual goal. The first part deals with updating sectoral policies and solutions related to new and increasingly dynamic risks facing the country. The second part works on additional linking of actions, exchange of information, and support between different institutions. Currently, the CIP area is not a development priority. However, ongoing activities certainly represent the basis for introducing the CI concept in Armenian legislation and practice. In Azerbaijan there is no normatively or organizationally established CIP framework. The greatest attention is paid to the protection of energy infrastructure and it is likely that this approach will be the nucleus of building a future system of protection of the entire national critical infrastructure.

Transatlantic cooperation partners, the US and Canada, have implemented strategies, laws, systems, sectoral policies, and above all, cooperation between various stakeholders in CIP. The United States identifies 16 CI sectors, Canada 10. Both countries in each sector have developed cooperation of all key partners – from the federal to the local level with special emphasis on cooperation with the private sector that owns and manages most national CI in both countries. The approach in considering the challenges to the protection and functioning of CI is an all-hazard approach, which includes developed concepts of cooperation and exchange of information between all relevant participants. The US and Canada have highly developed solutions in CI physical security and cybersecurity. As individual infrastructures are interconnected in both countries (in the energy and transport sectors), countries work together to strengthen resilience and protect CI. Also, together with Great Britain, Australia, and New Zealand, they form the Critical Five group, which, among other things, work on forging a common understanding for CI (these are countries that globally have the most developed principles and practices of CIP). The examples of solutions that the USA and Canada have are a very good basis for inspiring all countries to regulate the CIP area.

19.3. Public-Private Partnership

Public-private partnership is essential because a significant number of national CI is privately owned. In the West (Western Europe, USA, and Canada), it is estimated that up to 85% of the national CI is privately owned, so cooperation between the public and private sectors is needed in all areas important to strengthen resilience and protect CI. The situation in the countries covered in this book it will be shown in the analysis below. The analysis will be conducted, as in the previous subchapter, at the group level. For this and the next subchapter, it is necessary to state that there are fewer real indicators and examples of good practice; therefore, the mentioned parts are shorter than the previous ones.

Regarding the public-private partnership, the European Union has made the necessary steps in order to further increase it, since it's not always widely considered or properly implemented in the Member States. One step in this direction is constituted by the combined effort of the proposals – published on the 16th of December 2020 – for new directives respectively on *Network and Information Systems*[1] (aimed at amending the *NIS Directive 1148/2016*) and on the *Resilience of Critical Entities*[2] (CER Directive) which include specific measures aimed at facilitating the interaction between the public and private entities and the maximization of their efforts in protecting CI and essential services. The text of the so-called NIS 2.0 enhances the role of the Cooperation Group in shaping strategic policy decisions on emerging technologies and new trends and increases information sharing and cooperation. At the same time, it proposes the establishment of the Cyber Crisis Liaison Organisation Network (EU-CyCLONe) to better coordinate and manage large-scale cyber crises. The synergy among these two proposals is also clearly visible in article 16 of the CER Directive which establishes the Critical Entities resilience Group and explicitly foresees that the "The Critical Entities Resilience Group shall meet regularly and at least once a year with the Cooperation Group established under [the NIS 2 Directive] to promote strategic cooperation and exchange of information". Even though the aforementioned initiatives look a little more aimed at the public-public cooperation (among the Member States), the Commission is doing an unprecedented effort in bringing the public-private partnership to the stage of evolution it deserves to face the current challenges. An important initiative in this direction is constituted by the upcoming establishment of the European Cybersecurity Competence Network and Centre[3] whose mission is to establish and maintain communities composed of public and private entities in strategic sectors of cybersecurity in order to remove the fragmentation in the areas of research and industry and promote a better alignment among the actors of this playing field. The proposal for regulation "establishing the European Cybersecurity Industrial, Technology and Research Competence Centre and the Network of National Coordination Centres", which is in its later stage of approval by the Council and the Parliament and it's due to be officially promulgated by the end of 2021, has the objective to consolidate and improve the goals already foreseen in the 2016 *Commission Communication on Strengthening Europe's Cyber Resilience System and Fostering a Competitive and Innovative Cybersecurity Industry* by bringing together the research, industry and public sector communities in Europe. Thanks to the availability of funding to run specific projects and joint initiatives, it is foreseen that this regulation will contribute in a decisive manner in reducing the barriers and limitations therefore fostering a successful and impactful cooperation between public and private partners in the next future.

Croatia has indicated the importance of public-private partnerships in security strategies, the *Law on critical infrastructure*, the *Law on private protection*, and the *Law on public-private partnership*. Regardless of a strategic and normative stronghold, public-private partnerships have not been developed to the extent expected by the owners or managers of the national CI, primarily those coming from the private sector. All the necessary procedures and models for cooperation and exchange of information have not been established, and cooperation is not required. However, cooperation exists between various institutions from the public and private sectors (for which certain examples of good practice are provided in the chapter on Croatia), but there is certainly room for significant improvements.

Privatization of infrastructure is a rising trend in Turkey in a variety of different sectors. Energy is one of the most prominent sectors in which the private sector's role is rapidly increasing. The situation is similar in the health sector, where the private sector operates many hospitals. In such a context, significant steps have been taken to ensure the private sector's engagement in the CIP area. However, an institutionalized/specialized public-private partnership that holistically focuses on CIP is still absent. Although Turkey has the most developed approaches of all EU Enlargement Agenda countries and the largest number of institutions and partners involved in the CIP area in the public-private partnership part, the status quo in developing this part is visible. When it comes to public-private partnerships in BiH in the function of CIP, there is no uniform strategic and normative framework. Despite the nonexistence of a uniform state framework, cooperation among potential CI subjects and private security companies is present in BiH. Mainly, the owners or managers of CI rely primarily on their own resources in all protection segments. At the same time, the state takes care of state security issues and then indirectly on the security of CI. Serbia, like the previously mentioned example of Croatia, has enacted *Law on Public-Private Partnership and Concessions*. Still, this Law does not recognize the term "critical infrastructure"; however, it is clear that many facilities and services of public interest mentioned in Law are part of the overall CIP area. This will need to be harmonized when establishing an overall CIP system and a public-private partnership in CIP. The main development and implementation challenges of the future public-private partnership framework include efficient cooperation, the process of exchanging information and good practices; and establishing appropriate channels of communications. Currently, private security companies are very active, encouraging issues of standardization and the establishment of appropriate procedures and a common approach to all open issues. Montenegro has a very similar situation as all Western Balkans countries where a certain number national of CI are still majority state-owned, which in part facilitates public-private partnerships in the CIP field as the state is therefore obliged to protect its own CI. The chapter on Montenegro shows through a case study the protection of CI in the energy and transport sectors. Like all other countries within this analytical bloc, Montenegro has to work in the establishment and development of public-private partnerships in the protection of CI. In recent years, public-private partnerships have become a key initiative in the Republic of North Macedonia and have gained an increasingly prominent role. Each entity contributes to the system's specific resources, participates in planning, decision-making, and strives to enhance the resilience and protection of CI. The private sector is often the owner and manager of many national CI like communications, the IT sector, and transport facilities. The Macedonian normative model of public-private partnership is regulated by the *Law on Concessions and Public-Private Partnership* and *Law on Securing Persons and Property.* However, due to the lack of a systemic law on CI, the entire topic is approached in a sectoral manner and without the possibility of resolving vital parts of the CIP public-private partnership, such as exchange of sensitive information and data, as well as an exchange of knowledge and experience. Albania is currently in the position that public-private partnership is essential for it in terms of efforts to improve and develop the infrastructure and the provision of public services that promote economic growth and a source of alternative funding. The value of public-private partnerships in CIP has been recognized in the next areas: energy, transport, and information networks (internet) security. In these areas, steps and administrative actions are being taken to best respond to various stakeholders' needs and requirements. As the youngest state of all analyzed and with significant challenges described earlier, Kosova has the most challenges in the field of public-private partnership in the protection of CI, accordingly. In Kosova's case, the lack of experience and limited implementation capacities are present and followed by the lack of well-defined and formal cooperation between the CIP public and private sectors and actors. Aware of the shortcomings and challenges, Kosova has determined that building a public-private partnership in security (clearly including the CIP area) is a strategic goal of the state.

As mentioned in the previously analyzed section, covering Turkey, Ukraine has the largest activity within the public-private partnership area in this group of countries. The *Law of public-private partnership* was enacted in 2010 to enable private investment in public infrastructure but did not refer to CIP. Currently, the protection of CI through public-private partnerships is implemented in two directions. In the first, which refers to physical infrastructure, the cooperation is reflected in joint working groups in CI sectors where these working groups work on identifying CI, risks threatening these infrastructures, and proposing improvements in legislation. In the second, a more developed form of cooperation exists in the field of cybersecurity. Where different public and private actors work together to protect national CI. Nevertheless, in general, Ukraine still needs a proper regulatory basis for public-private partnership in the CIP with the description of the private and state agencies' mutual obligations concerning the protection of CI. Regarding Belarus, which has a solid legislative framework regarding public-private partnerships, the actual concept of public-private partnership has only recently entered the political and legal lexicon in the country, and none of the existing normative acts make any references to CI. Belarus is working intensively to provide the best possible basis for the development of public-private partnerships. It seeks to attract domestic and foreign investment in the development and upgrading of infrastructure. *National Infrastructure Strategy for 2016–2030* (approved in 2015, updated in 2017) is focused on public-private partnership projects in social, energy, and transport infrastructure. The strategy lists 100 key projects for possible implementation of a public-private partnership mechanism. Although none of these projects is related to CI, experience from other countries in implementing public-private partnerships in infrastructure speaks in favor of building knowledge and procedures that will enable Belarus to apply the same model in the CIP area in the future. In Moldova, there is a cooperation between the public and private sectors in the protection of CI, this refers, among other things, to the fulfillment of the obligation of anti-terror legislation, and to some extent, the public and private sectors have to align also with the defense rationality of the state. It is interesting how Moldova, through *the Law on Public Property, privatization of the goods of public property*, has determined the number of categories in which facilities, networks, and systems are excluded from the privatization. Such practice has not been established so far, as evident in other analyzed countries. In Georgia, the normative model of public-private partnership is regulated by the *Law on public-private partnership* implemented in 2018. Law determines the legal basis for public-private partnerships, including the rules and procedures related to the development and implementation of public-private partnership projects, the principles of public-private partnership, and relevant institutional systems, as well as other matters related to public-private partnerships. Unfortunately, this Law does not deal with public-private partnerships within the field of CIP. However, there is a realization that the public-private partnership in the CIP area is significant. There will likely be a development of the normative framework to regulate the aforementioned in the near future. Armenia is an example of a country where a state-centric approach is still expressed in protecting CI, both public and private. This is gradually changing; Armenia pursues developing and applying mechanisms for constructive cooperation between the public-private sector as a type of new culture. This is currently most visible in the field of cybersecurity, which in the next step can serve as an example of good practice for other areas as well. In Azerbaijan, the government is making efforts to promote public-private partnerships and provide the private sector with better business conditions. Specifically, these are the following activities: the government has stepped up administrative requirements, expanded e-government services and implemented a number of policy initiatives aimed at promoting the private sector.

The protection of CI in the US is a shared responsibility between private sector owners in each sector and governmental agencies at the federal, state, local, tribal, and territorial levels. The US government, in coordination and collaboration with key stakeholders, has developed a number of standardization documents that provide structure to protection programs across the 16 sectors. The government and the private sector each have unique responsibilities, and the perspectives of each are equally important. It is significant to emphasize that cooperation between the public and private sectors is voluntary. This voluntary collaboration among sector stakeholders is a key element in the overall *National Critical Infrastructure Protection Plan* and is vital to successful CIP efforts. As part of the Plan, public and private sector partners in each of the 16 CI sectors and the state, local, tribal, and territorial government community have developed sector-specific plans that focus on the unique operating conditions and risk landscape within each unique but interdependent sector. The proliferation of cyber-attacks necessitated a new approach toward synchronizing action, collaborating on public-private efforts, and coordinating intelligence. In 2018, to better address the growing complexity of protecting the nation's CI and centralize historically decentralized efforts, the *Cybersecurity and Infrastructure Security Agency Act* established the CISA. CISA results from a collaborative effort to identify gaps in communication and coordination and develop a method to better address areas of operational weakness. It is an example of how governments must continuously assess structure and oversight and be willing to adapt and adjust to meet changing operational and strategic environments. The United States is a leading example of how public-private partnerships in the field of CIP can and should be regulated. Regarding Canada, more than 85% of Canada's infrastructure is owned and operated by the private sector. Therefore, strengthening the resilience and protection of CI requires public-private partnerships among all partners and stakeholders. Public Safety Canada (CIP-responsible federal authority) works closely with federal departments and agencies, provinces and territories, private sectors, and international counterparts to build partnerships. Initiatives are focused on building, sustaining, and enhancing collaboration with all partners within the CI community. Moreover, Public Safety Canada also recently launched the Cyber Security Cooperation Program as a means to improve the security of Canada's vital cyber systems. Furthermore, Canada has invested to promote a coordinated, global approach to CI. As an example, Canada and the US have achieved great results toward meeting shared objectives of enhancing the resilience of CI through the *Canada–United States Action Plan for Critical Infrastructure,* and both countries are actively working within the Critical Five (mentioned in the previous subchapter) with other countries that have the same highly developed solutions in the CIP area.

19.4. Cooperation with European Commission

The European Union, especially European Commission, is focused on the development of different cooperation platforms of Member States, EU Enlargement Agenda and EU EaP countries, and Transatlantic partners. The Commission is also cooperating, and endeavoring to cooperate, with owners or managers of CI, international organizations, and interested experts. Cooperation between the Commission and the various partners is based on the following models: Member States national contact points meetings, Meetings with EU Enlargement Agenda and EU EaP countries and Transatlantic partners, Cooperation through ERNCIP and CIWIN, and EU-funded projects. The Republic of Croatia actively participates in a formal way (format) of cooperation, which is defined and provided by the

EPCIP. Special emphasis is on ECI protection contact points' meetings as a way of achieving a common understanding of the challenges and threats to CI as well as solutions to strengthen resilience and protection – among all Member States. In addition, Croatia has used the opportunity of European Commission Funds and implemented the project that helped improve knowledge and provide a better understanding of CIP, especially regarding national public policies.

Turkey has a long history of relations with the European Union and has the longest candidate country status in the EU Enlargement Agenda. In that sense, as a candidate country, adaptation and alignment of most of Turkey's national legislation is necessary. Throughout this process, the CIP topic is covered with the civil protection part, which takes place under the Environment Chapter. In connection with the above, it should be noted that the *Critical infrastructure protection Road Map* was developed as a guideline to align Turkish CIP policies implementation with those of the Union. Within the mentioned document, a list of obligations was made: preparation of a draft bill for alignment with EU regulations, closer cooperation with relevant EU stakeholders, and integration of Turkey's practices with EU's CIWIN. Nevertheless, so far, most of these action items are not yet realized. Turkey is currently working to align its legislation with the provisions of the *NIS Directive*. To this end, it has carried out several projects aimed at assisting in this alignment. Besides, it seeks to be active in scientific and research projects organized by the Union and be associated with various forms of cooperation that take place under the auspices of the European Union. BiH so far have no recorded history of cooperation with the European Union in general, or the European Commission in particular, in the field of CIP. Cooperation between Serbia and the European Union is currently on two tracks. The first, under the *Law on critical infrastructures* into which it has transposed *Council Directive 114/08/EC*, considers the state of ECI on its own territory and potentially significant CI on the territory of the EU Member States. Accordingly, the next step will be communication and coordination with neighboring countries and the European Commission in this regard. On the second track, Serbia participates in several EU-funded projects that, to a greater or lesser extent, have direct and indirect contact with CIP issues and thus strengthen its own capacities. Montenegro is in the same position as Serbia, has passed the *Law on the determination and protection of critical infrastructure of Montenegro* by establishing its own system of CIP and considers issues of identification and designation of ECI in its territory and the territory of other countries. Montenegro has been in negotiations with the EU since 2012. During that process, many laws are aligned with different EU directives that have some significance in the wider context of different sectors in which it is possible to identify national and ECI. North Macedonia has similar challenges like BiH, if it did not pass the *Law on critical infrastructure*, which would serve as the initial basis for developing the CIP area and a stronghold for cooperation with the European Union. In the absence of the aforementioned, North Macedonia seeks to be present at meetings organized by the

Union with external partners on CIP issues. In the broader context of the discussion on this topic, in December 2018, North Macedonia was formally applied to participate in the Hybrid Threat Survey for the EU to identify their key weaknesses and strengthen resilience to hybrid threats. The knowledge gained through this activity certainly had a positive effect on the perception of challenges in protecting CI. Albania is very active in regional and European projects and programs related to the energy and transport sector, so it has transposed all relevant EU directives in these sectors, which are also indirectly related to the CIP area. Regarding direct cooperation with the European Commission within the CIP field, Albania participates in CIP meetings and workshops that the Commission organizes with external partners (as part of the European Program for Critical Infrastructure Protection) for deepening possibilities of cooperation. So far, Kosovo does not have any significant cooperation with the Commission in the field of CIP. The only thing they could achieve is that Kosovo's representatives were present at several CIP meetings organized by the Commission with external partner countries.

Ukraine has cooperation with the European Union in the CIP area on several levels. The first level is the general one, for which it can be stated that the CIP area is not included in the priorities of EU–Ukraine cooperation, and Ukraine still needs legislation that will initiate such cooperation. At the next level of observation, which is more specific and related to the energy and transport sector, Ukraine has particularly close links with the energy and transport infrastructure of the EU Member States due to its geographical location. In this area, joint measures and activities are being implemented between Ukraine and individual EU countries on different CIP aspects. There are well-developed forms of cooperation on securing the steady functioning of the Ukrainian gas transportation system, which could be designed as the ECI according to the understanding of the spirit of *Council Directive 2008/114/EC*. The next level of cooperation between Ukraine and the European Union is related to cybersecurity. The cybersecurity development of capabilities in combating cyber and hybrid threats is at the center of attention of EU–Ukraine cooperation. Here, Ukraine is developing a sectoral cybersecurity system according to the latest EU regulations and recommendations on cybersecurity measures in the energy sector. Additionally, the EU assists Ukraine through its EUAM to Ukraine. The EUAM allocated more than 2.5 million euros to various projects helping Ukraine in cybersecurity. The mission promotes the improvement of Ukrainian law enforcement agencies' technical equipment, conducts training, exchanges of experience, and panel discussions. Due to the European Union's position on respect for human rights in the country, Belarus has a reduced opportunity to cooperate with the Union's institutions, and thus within the CIP area. However, Belarus has been actively taking part in the multilateral formats of the EaP. Different forms of cooperation between Belarus and EU institutions still occur in cybersecurity matters, CI-related legal and technical capacity-building, and financial assistance. This cooperation is most pronounced through the participation of Belarus in various projects funded by the Union and concerning support partner countries in adopting EU standards, rules, certification, and accreditation in the gas and oil sectors, strengthening partner countries' capacity to respond to cybercrime and harmonizing legislation in the cyber domain, capacity-building support in the field of civil protection, and disaster management. Cooperation between Moldova and the European Union has no direct link with the protection of CI. However, it is oriented toward the traditional sector-based cooperation mostly in digital transformation and the energy sector. Moldova participates in available EaP states' programs funded and supported by the European Union, such as the EU4Digital or EU4Energy. Through these programs, the Union bilaterally supports the development of Moldova's state policies, in legal or technical terms, in the field of sectoral infrastructures. These programs are contributing to the CI, even if in a more limited format. Georgia has determined that the improvement and development of its own normative framework will be directly linked to harmonizing the country's legal system with the EU *Acquis Communautaire*. This also applies to the area of CI. In this regard, several projects are currently being implemented in Georgia to strengthen various areas of CIP and strengthen cybersecurity capabilities. In this significant effort, international partners (EU, NATO, US) are engaged in providing support for the CI reform. Like all EaP countries, Armenia has developed a certain form of cooperation with the European Union. However, so far, the area of CI has not been covered by that cooperation. Therefore, soon, Armenia will construct a deeper relationship with the EU in more cooperation areas than before, including CIP. The EU and Azerbaijan have established cooperation in several areas, such as trade and EU investment in certain sectors in Azerbaijan. However, no cooperation has been established in the CI area so far.

Cooperation between the US and Canada on the one hand and the European Union on the other is taking place in the format of regular meetings established as a way of exchanging experiences and best practices in CIP. EU–USA–Canada CIP expert meetings are covering various areas, from physical to cyber to hybrid threats through discussion on solutions and further development of cooperation in dealing with global challenges.

19.5. Education and Training

The European Union invests significant resources in education and training in the CIP area. The Republic of Croatia currently does not have a specific comprehensive program at higher education institutions, although there are some examples (such as the University of Applied Sciences) of courses covering CI concept and management. They are only intended for university students, without the possibility that the general public or public servants participate. Of course, there are some exceptions, such as 2014 and 2015 examples of seminars on risk analysis and risk assessment and management for stakeholders engaged in national CIP activities. In 2021, a program will be developed, and the obligation of education to all stakeholders will be prescribed. The same is valid for developing a model for the training of all key actors who are engaged in the CIP system, based on good practice of annual exercises that the Republic of Croatia conducted in 2018, 2019, and 2020 as simulation exercises based on cyber-attacks scenarios.

Turkey is the largest and most powerful country within the EU Enlargement Agenda Framework. It also has the largest number of universities, research centers, and laboratories. We can freely say that it has by far the largest opportunities for conducting research, education, and training in the field of CI. Nevertheless, the current situation is such that as in many other countries around the world, and within this observation framework, in Turkey, the number and level of academic and professional CIP course programs are insufficient. One of the biggest challenges that impede the development of widespread educational programs and the emergence of CIP as a formal field of study is underestimating and lack of interest since most of the existing efforts have not yet captured educators' and organizations' attention. There are, however, bright examples. For instance, Kadir Has University's Center for Cybersecurity and Critical Infrastructure Protection (CCIP) is one of the first established centers focusing on CIP research, such as security of ICS, energy, and other CIs. Other university programs and non-institutional courses aimed at training and coaching professionals are predominantly oriented toward cybersecurity. Turkish private cybersecurity companies and some state-supported institutions started to develop cybersecurity curricula and establish academies to support human capital development and training. BiH's situation is somewhat similar but clear as for any other observed country within this group; there are fewer opportunities than in Turkey's case. In BiH, over the recent years, the public, private and academic sectors have increasingly identified the need to develop a platform for CIP and cybersecurity. In that context, certain universities incorporated the modules to study CI and cybersecurity and have implemented significant cooperation with the state institutions. In addition to the university level, certain forms of education and training are also provided at the level of private agencies. Also, BiH, through cooperation with international partners, participates in various exercises aimed at capacity-building to manage CIP and cybersecurity issues. Serbia's situation is quite similar, where Serbia draws its additional experiences from EU-funded CIP projects in which it participates and from which it gathers experiences related to the entire topic of this area. One such project is described in more detail in earlier chapters. In Montenegro, several faculties offer academic education in their programs which indirectly deals with the issues of protection of CIs, such as the Faculty of International Economics, Finance and Business; Faculty for Information Systems and Technologies; Faculty of Polytechnics; Faculty for Food Technology, Food Safety and Ecology; and Faculty of Humanistic Studies – Department for Security Studies. The only program for Security Studies in Montenegro exists at the University of Donja Gorica (Podgorica). Academic disciplines and study areas that indirectly teach on CIP at the Bachelor level include national security, cybersecurity, security culture, crisis management, and international terrorism. Also, at the postgraduate level, there is a cybersecurity program, which directly teaches students on IT CI. An informal education program exists at the UPTC, which trains persons who intend to work specifically on protection tasks, following the *Law on protection of persons and objects*. Regarding CIP training activities, certainly, not many of these activities have been realized so far. North Macedonia has a modest CIP experience. It is at the beginning of developing a CIP system and (therefore) does not have a specific CIP program in any higher education institution. A subject entitled Critical Infrastructure is being studied only at the University of Ss. Cyril and Methodius, at the Institute of Security, Defense and Peace at the Faculty of Philosophy in Skopje. In addition, the scientific and expert public has initiated several conferences, seminars, training, workshops, and exercises related to CIP. The Private Security Chamber, which organizes these activities, is very active in the aforementioned. In Albania, regardless of an applicable legal framework, there is still no specific comprehensive education program and training in higher education institutions that can be linked to the CIP. So far, the only opportunity to acquire knowledge has been through attending various seminars, workshops, and exercises organized by international organizations such as NATO and the OSCE. Of all the countries analyzed, Kosova has the least recorded activities related to education and training in CIP. Kosova has received international assistance in drafting the *Law on critical infrastructure* and other documents. In these processes, EU institutions and other international organizations present in Kosova have provided basic training and workshops for public institutions and the private sector to understand the CIP framework better and prepare domestic experts to implement the Law.

In the EaP states group, the situation is quite similar to the previous group – the largest country (specifically, Ukraine) has the most opportunities and most activities in education and training in the CIP area. Ukraine has well-developed education and training system in response to natural and technogenic threats that include CIP issues. There are some universities and research institutes that have these kinds of programs. Additionally, in the last decade, Ukraine established a number of educational programs in information security, information security management, and cybersecurity. Currently, they provide not only education services but also research and training in all mentioned fields. The CIP issues were introduced in a curriculum of the universities that provide educational and training programs for the security and defense sector of Ukraine. Depending on specificities, CIP modules are included in counterintelligence, law enforcement, emergency, cybersecurity programs. In addition, since 2015 in Ukraine was organized a series of events to support the CIP system's implementation in a form of workshops and seminars discussing the problem with public servants and experts, and they have been conducted by the NISS. Ukrainian ministries and governmental agencies held expert meetings and conferences discussing the CIP. Ukrainian efforts in raising awareness and education of some ministries have been supported through international cooperation programs. National-level TTX on the issue of resilience of CI are also conducted. The target audience of exercises is the personnel of government agencies and ministries in the field of energy, emergency service, national security that include energy government officials, military personnel, national police and other institutions and agencies responsible for the protection and building resilience of electricity supply by improving plans, procedures and processes at a national level. In Belarus, the current situation is that responsibility for developing CIP competencies and training is a joint obligation of state and non-state institutions. One of the key highest educational institutions in Belarus which provide CIP education is the University of Civil Protection. The University also offers training for experts in emergency response and technosphere security, industrial and radiation safety, civil defense, and protection from natural and man-made character emergencies. The university has served as host to many regional and international practical conferences, expert talks, and joint training with international organizations. Besides, each CI operator has a legal obligation to conduct various forms of education and training of its employees, which they do. In Moldova, there is no systemized format of education or training in the CI field that involves the public servants and institutions. In terms of cyber-related infrastructure, the Public Administration Academy that prepares public servants to serve the government's positions conducts courses that are annual state's requests. Additional opportunities for acquiring the necessary knowledge are mainly related to certain courses and workshops conducted by individual official state institutions for their employees or participation in international projects and programs such as those conducted by the European Union, NATO, and the US. Everything related to the CIP area is a fairly new topic in Georgia. Therefore, there is limited knowledge within the organizations on CI management security of their respective sectors. Until now, Georgia does not have a specific program in any higher education institution within the field of CIP. In Armenia, there is a developed awareness within the academic sector and various public and private institutions about the importance of CI. In the formal part of education, CIP topics are not directly introduced as independent disciplines, but they are taught in a wider range of security, management, and crisis management. Also, non-formal forms of education and training involving CI operators and managers, government services, and citizens are organized and implemented. Azerbaijan does not have established curricula within the education system that covers the area of critical infrastructure, but the training comprises of a number of occasional workshops, courses and exercises in which domestic experts in cooperation with international organizations have the opportunity to exchange knowledge and experiences.

The largest number of formal and informal forms of education and training are in the US and Canada. We can freely say that the US is a global leader in this area. Education and training in CIP in the US are extensive and widely available. The main source of education and training for CIP is the Department of Homeland Security, consisting of direct delivery and identified training and education providers, supported through Department funding. In addition to the above, numerous academic and professional programs, courses, and workshops provide ample opportunities for all those interested in acquiring knowledge and skills in the CIP field.

The possibilities are exceptional and very accessible to different levels and for different needs. In Canada, based on the data found in the chapter on Canada, we see an extensive network of opportunities provided by the Public Safety Canada (CIP federal responsible authority) for the protection of physical CI and their cyber protection.

Chapter Conclusion

The research showed that all countries understand the importance of strengthening resilience and protecting CI in its physical form and its cyber dimensions. CI is extremely important for each of the analyzed countries, regardless of whether it had an official definition, a certain concept, drafted the *Law on critical infrastructure protection*, and established a system of its protection. The foundation of CI protection and resilience is in the development of policies, strategies, and action plans, and achieving an adequate security level is a complex task that requires multilevel involvement and mechanisms. Likewise, CI is equally important for international organizations such as the European Union, which is making great efforts to harmonize its Member States' actions and it is offering support to the neighboring countries in various ways, including the exchange of experiences and best practices with transatlantic partners. For the Member States, it is valuable to adopt coherent EU public policies for achieving coordinated cross-sectoral action and trans-sectoral cooperation mechanisms that can ensure better security through the whole community. Non-EU countries are also aiming to achieve a higher level of security for national assets, networks, and systems that are the most important to the state's functioning, in line with their capacities and capabilities. Through the research, we learned about different approaches to the CIP area. We realized that financial resources are essential but not crucial in implementing activities that will develop a strategic and normative framework, draft bylaws, cooperate with various stakeholders, cooperate with the European Union on the harmonization of national implementations, and, last but not least, on education and training. We recognize that one of the most important elements of the CIP equation is the human factor that should ensure the vision and direction of developing this area in the respective countries and organizations. The analysis also confirmed the empirical findings that countries with greater opportunities have broader and better solutions in the CIP area. This is partly determined by the fact that these more advanced countries have dealt with this area in a very focused manner for many years. So, the research showed countries at different levels of opportunities, development, and opportunities. Smaller countries or countries with fewer opportunities can make up for their shortcomings by learning and collaborating with those who are currently more advanced than them. This book provides an excellent opportunity to have an overview of the state of art and influence the decision-making accordingly. We have seen that many countries use the opportunities to exchange knowledge and experience, participate in international projects and programs, and special frameworks of cooperation with the European Union – to overcome the gaps in CIP. The Union is a gold mine of opportunities; it has an open door for everyone, and partners need to ask for help and assistance. Because ultimately, acting locally, all partners create safer critical infrastructures, which in turn serve as platforms for all of us in our maintenance of existing systems, development of new capabilities, and improvement of processes of all forms.

The latest bits that the European Commission has just put on the table should not send a wrong message to the countries that are developing their national frameworks and international cooperation. These countries surely have to look at the latest developments but need to make sure that their baseline and basic pillars are correctly established and have reached a good level of operational maturity. Topics like cybersecurity certification of products, services and processes, hybrid threats, and critical entity resilience[4] are to be duly taken into account but need to be implemented in an established framework. The risk, otherwise, is that countries may underestimate threats and vulnerabilities and fail in embracing an all-hazard approach.

Directives like the Seveso series and the one on ECI – which is about to be disbanded – still constitute a very solid reference for countries that are in the phase of improving their level of maturity in dealing with the security and continuity of critical infrastructures and essential services. Concepts like (1) prioritization of security measures, (2) risk analysis which embraces health, safety, security, and environment, (3) the assessment of transboundary externalities, (4) prevention and preparedness, just to name a few, shouldn't be missing in any of the national frameworks established in the last 20 years (from 9/11 and onwards).

Notes

1. European Commission "Proposal for a Directive of the European Parliament and of the Council on measures for a high common level of cybersecurity across the Union, repealing Directive (EU) 2016/1148", COM (2020) 829 final, 16th of December 2020, available at: https://ec.europa.eu/newsroom/dae/document.cfm?doc_id=72166.
2. European Commission "Proposal for a Directive of the European Parliament and of the Council on the Resilience of Critical Entities", COM (2020) 823 final, 16th of December 2020, available at: https://ec.europa.eu/home-affairs/sites/default/files/pdf/15122020_proposal_directive_resilience_critical_entities_com-2020-829_en.pdf.
3. See https://digital-strategy.ec.europa.eu/en/policies/european-cybersecurity-competence-network-and-centre.
4. Reference is made respectively to the Regulation (EU) 2019/881 (Cybersecurity Act), "the landscape of Hybrid Threats: A Conceptual Model", European Commission's Joint Research Centre, 2021 (available at: https://publications.jrc.ec.europa.eu/repository/handle/JRC123305) and the Proposal for a Directive on the resilience of entities, European Commission, 16th of December 2021 (available at: https://ec.europa.eu/home-affairs/news/commission-proposes-new-directive-enhance-resilience-critical-entities-providing-essential_en).

Chapter 20

Conclusion

Alessandro Lazari and Robert Mikac

This book is a three-year journey that was extremely interesting, dynamic, and sometimes unpredictable. Its concept was drafted based on the recognized need and the clear intention of shedding some light on the frameworks of neighboring countries of the EU and with the ultimate goal of developing a sort of reference manual that should foster awareness regarding the state of the art and the challenges but even more importantly, facilitate effective and impactful cooperation within local, regional, continental and international dimensions. Much time has passed from the initial idea to the realization, a lot of conversations, correspondence, looking for co-authors, including some of them giving up. The process itself took longer than expected, especially since during its preparation, the whole world has been affected by the COVID-19 pandemic and the resulting crisis that has temporarily put the focus of most of us on other important life and family responsibilities. The journey was ultimately successful for two key reasons, the first being the editorial team at Routledge Francis & Taylor, who followed us all the time and gave us support and advice on designing the book. Another reason is the project team, which comprises exceptional individuals and experts ranging from a newly graduated in political science to a Minister of Defense. The impressive biographies of the contributors to this book are represented at the end of this publication and constitute also a landmark of how multidisciplinary the CIP dimension is and should be. Cooperation with everyone on this project was extremely successful and useful, and we all learned a lot from each other, which is the primary goal of any – even embryonic – form of cooperation. Therefore, we believe that also non-specialized readers will find interest and significant elements within the pages of this publication.

In order to provide the readers with a full view of the state of the art of CIP, we have divided the book into five units. The first unit is an introductory part, which contained an inception to the methodology and driving principles taken

DOI: 10.4324/9781003273769-25

into account in the process of drafting the book. The second chapter describes the European Union Critical Infrastructure External Dimension's current position: from neighboring policy to Transatlantic cooperation. The third chapter is an analysis of the current reach of the Republic of Croatia as the country that most recently has joined the European Union and therefore has been taken as a reference model for countries that are in the process of becoming full members of the Union or that are aiming at establishing their national CIP frameworks. The second unit presents an analysis of strengthening resilience and critical infrastructure protection from the side of seven EU Enlargement Agenda states (Turkey, BiH, Serbia, Montenegro, North Macedonia, Albania, and Kosova). The third unit offers an analysis of the frameworks and efforts of six EU EaP states (Ukraine, Belarus, Moldova, Georgia, Armenia, and Azerbaijan). Many similarities have been observed between the solutions applied and pursued by these 13 states. The fourth unit brings an analysis of Transatlantic cooperation partners (USA and Canada) and the European Union. These two countries have far more capacity and comprehensive CIP solutions than any other country therefore they were included since they offer significant lessons learned and many potential means of tackling CIP challenges for countries that are in the phase of defining or fine-tuning their national frameworks. The last, fifth section, provides a comprehensive analysis of the overall research. Besides, each chapter, which was a case study analysis of each country, was divided into five subchapters with the following objectives: (1) Introduction – a description of the current state of development of the CIP area with special focus on whether there is any form of normative inspiration in national legislation or direct reference to the European Program for Critical Infrastructure Protection policy package (including the *Council Directive2008/114/EC*) or the Cybersecurity Policy package (including the *Directive (EU) 1148/2016*). (2) Implementation – In this part, the national CIP governance was represented. That included the structure, stakeholders, national plans, and overall critical infrastructure resilience and protection system. (3) Public-private partnership – here was described of public-private partnerships established in the country and how they work and are regulated. (4) Cooperation with the European Commission – this section brings cooperation with the European Union as a whole, and with the Commission in particular. (5) Education and training – formal and informal forms of academic and professional training in the CIP field are analyzed here. This structure within all chapters allowed us to consider all the important items for this research and the possibility of a comparative comparison of CIP solutions in different countries.

The last part of the conclusion of this challenging publication finally encloses the following recommendations and considerations that address, as final recipients, the entire community of stakeholders involved in CIPR; from academia to subject matter experts and from operators of CIs to staff of local, regional, national and international public bodies.

These final considerations are intended as some of the elements to eventually take into account when tailoring down projects, activities, training and cooperation frameworks in the field of CIPR. It is the opinion of the authors that not all of them are always applicable to every context or case, since the discipline of Critical Infrastructure Protection is always influenced by political, legislative, economic, cultural, historical and geographical reasons. However, there's the feeling and belief that many of the following considerations, together with state of the art represented in the previous chapters, could be very useful in order to assess potential ways toward and also "road ends" in CIPR, at all levels, from the most strategic to the most tactical.

The sections regarding the "cooperation with the EU or the European Commission" clearly show that the EU is considered a trusted partner and its respective policies and regulations are always taken into consideration also from the countries that are not part of the enlargement agenda. For these reasons the Union should keep planning joint efforts with the neighboring countries, pending that it's a fact that such countries are crossed by or host infrastructures of European relevance (e.g., the Trans Anatolian and Trans Adriatic pipelines – TANAP & TAP) from which the continuity of vital services are heavily dependent.

The *external dimension* has been and still nowadays is among the most successful and impactful elements of the EPCIP. Unfortunately, this part of the program has not reached the countries of North Africa (most probably because of the difficulties in establishing contacts with the correct PoC for CIP), which are also neighbors to the Union and very often host infrastructures of European relevance (mainly in the Oil and Gas sector). On this point, the Commission should foster an improved version of the EPCIP with an updated list of countries to reach, with the aim to establish cooperation frameworks that should foster a higher level of CIP maturity at both governmental and operational levels to arise.

The Achilles' heel of security is still represented by the difficulties encountered by public and private parties in establishing secure information sharing for CIP and Cybersecurity. On this point, further efforts should be planned by all stakeholder in view to find a way to make important information available in due time (e.g. early warnings). The – still ongoing – COVID-19 pandemic has taught several important lessons in the areas of prevention, preparedness, and response. However, the full potentiality of such areas cannot be unleashed in case of lack of information sharing the mechanism that are the enablers and triggers of decision making in all sorts of cases and scenarios, from the lighter to the most severe one.

Finally, the analysis of the national variations of the countries considered by this publication, provides a set of scenarios and lessons learned that could be a useful starting point for countries that are in the phase of establishing or improving their respective frameworks. Since the domain of CIP is characterized by many maturity-assessment models, for the purpose of this publication, a basic set of key performance indicators for the establishment of national CIP framework, could be the ones described in the following.

A National CIP framework should include, at least, an unambiguous definition of "national critical infrastructures and respective sectors", a CIP strategy and means of national governance, together with procedures for the designation of national and international critical infrastructures. The most successful EU approaches (e.g., UK's CPNI), demonstrate that countries should also consider the establishment of a national center for protection of critical infrastructures or providing a mandate to an existing entity. The establishment of centers as such, which are normally provided with sufficient autonomy, creates the condition for an impactful and efficient interaction with operators of CIs, as well as room for the utilization of the soft law (e.g., guidelines, recommendations) to provide guidance to the operators of critical infrastructures. In this context, many countries have provided national operators with security frameworks of reference to be used for risk analysis and maturity assessment purposes.

The role of the reference security frameworks (e.g., NIST Cybersecurity) should not be underestimated in the establishment of joint procedures for the protection of critical infrastructures, because of their ability in fostering harmonization (including the provision of risk scenarios) and more if such tools are coupled with means of reporting the status of the "operator security plan" to the competent authorities. The availability of such reports, together with the establishment of incident-notification mechanism (e.g., the ones adopted in the NIS Directive) normally provides a very useful picture that can be embedded into National Risk Assessments (NRAs) to be shared back with the operators to provide them with improved awareness on their role in respect of national security.

Another important matter is the one related to the organization of local, national, and regional exercises, training and simulations to improve response in case of systemic incidents and crises. Even though the international CIP community is discussing "resilience" for a long time already, the impression is that countries are not investing enough on joint preparedness and coordination of responses. It is the opinion of the authors that given the complexity of modern legal frameworks and of the CIP domain as well, pilots and dry-runs should also be extended to new regulations and policy as well. The possibility to test and simulate the impact of new laws or procedures prior to their effective entry into force, seems of pivotal importance if the avoidance of the risk of duplication of duties, over bureaucratization and potential loss of security investments are on top of national security agendas.

Last but not least, it is important to draw attention to the matter of "transboundary externalities", as introduced and advocated by the EPCIP and the *Directive 2008/114/EC* on ECI. Since it's a fact that all countries are every day more and more interconnected and dependent one another, governments and competent authorities should consider fostering the organization of bilateral and multilateral workshop with neighboring countries in order to fine-tune the coordinated response to large scale events, including the identification of infrastructures of regional relevance and potential solidarity mechanisms. On these matters, in fact, the government should foster the establishment of regional fora with long-term agendas to make sure that neighboring cooperation is kept as a priority.

Every potential layer to be considered for building up or improving national frameworks has been described in this final chapter, can find a key enabler in the academia which should always play a very important role in the field of CIP. Spanning from comparative studies to more experimental research or applied research, academia can provide very qualified and effective support to both governments and operators in the protection and resilience of critical infrastructures. Among the areas of competence, it is worth highlighting the following domains:

- Definition of education and training courses to foster the emergence of qualified experts in the multidisciplinary areas of CIP;
- Provide governments and critical infrastructures with a top-notch solution to be applied in the field of security and resilience;
- Develop fictional scenarios to be used by governments and critical infrastructures to test themselves outside of their respective comfort zones and draw conclusions on potential attack vectors, threats, vulnerabilities and disruptive events to consider for further improving the protection and resilience of critical infrastructures.

As irrefutably demonstrated by all the arguments as well as by all the analysis of the previous chapters, the path to pursue to reach a mature model for the governance of critical infrastructure protection is still long and dense of challenges and difficulties to overcome. It is exactly for this reason that this publication has been compiled with the goal of shedding some light on the current state of play and therefore fostering more impact and less bureaucratization around the matter of cooperation and coordination in the areas of CIPR.

Finally, according to the authors, the focus of CIP should be still the making "any interruption or manipulation of critical infrastructures brief, infrequent, manageable, geographically isolated, and minimally detrimental to the welfare"[1] of a nation and its neighbors.

Note

1. This intent has been intentionally borrowed from the President of the United States of America as formulated in the PDD-63 of 1998 because of its comprehensive and clear definition, in the author's perspective, of what the real aim of CIP should still be.

References

SECOND CHAPTER (Introduction chapter)

Council of the European Union (2008), *Council Directive 2008/114/EC of 8 December 2008 on the Identification and Designation of European Critical Infrastructures and the Assessment of the Need to Improve Their Protection*, Brussels, 2008/114/EC, EUR-Lex, Official Journal, https://eur-lex.europa.eu/LexUriServ/LexUriServ.do?uri=OJ:L:2008:345:0075:0082:EN:PDF

Council of the European Union (2007), *The Prevention, Preparedness and Consequence Management of Terrorism and Other Security-Related Risks*, EUR-Lex, Official Journal, http://eur-lex.europa.eu/legal-content/EN/ALL/?uri=CELEX:32007D0124

Engdahl, Eva Maria (2016), *The European Programme for Critical Infrastructure Protection,* Gas Infrastructure Europe, https://www.gie.eu/index.php/events-diary/workshops/2016-workshops/gie-security-day-brussels-16-november/26303-critical-infrastructure-protection-in-europe/file

European Commission (2020), *EU Security Union Strategy*, https://eur-lex.europa.eu/legal-content/EN/TXT/?qid=1596452256370&uri=CELEX:52020DC0605

European Commission (2019), Cybersecurity, https://ec.europa.eu/digital-single-market/en/cyber-security

European Commission (2017), *Commission Staff Working Document on Assessment of the EU 2013 Cybersecurity Strategy*, EUR Lex, Official Journal, http://ec.europa.eu/transparency/regdoc/rep/other/SWD-2017-295-F1-EN-0–0.PDF

European Commission (2013), *Commission Staff Working Document on a New Approach to the European Programme for Critical Infrastructure Protection: Making European Critical Infrastructures More Secure,* EUR-Lex, Official Journal, https://ec.europa.eu/transparency/regdoc/rep/10102/2013/EN/10102-2013-318-EN-F1-1.PDF

European Commission (2006), *European Programme for Critical Infrastructure Protection,* EUR-Lex, Official Journal, http://eur-lex.europa.eu/legal-content/EN/ALL/?uri=CELEX:52006DC0786

European Commission (2005), *Green Paper on a European Programme for Critical Infrastructure Protection,* EUR-Lex, Official Journal, http://eur-lex.europa.eu/LexUriServ/LexUriServ.do?uri=CELEX:52005DC0576

European Commission (2004), *Communication on Critical Infrastructure Protection in the Fight Against Terrorism,* EUR-Lex, Official Journal, http://eur-lex.europa.eu/legal-content/EN/TXT/?uri=CELEX:52004DC0702

European Commission and High Representative of the European Union for Foreign Affairs and Security Policy (2013), *Cybersecurity Strategy of the European Union: An Open,*

Safe and Secure Cyberspace, EUR-Lex, Official Journal, http://eeas.europa.eu/archives/docs/policies/eu-cyber-security/cybsec_comm_en.pdf

European Council (2010), *Internal Security Strategy for the European Union: Towards a European Security Model,* Brussels, 25 and 26 March 2010, https://www.consilium.europa.eu/media/30753/qc3010313enc.pdf

European Council (2007), *Treaty of Lisbon Amending the Treaty on European Union and the Treaty Establishing the European Community,* Official Journal of the European Union, C 306, 17 December 2007, https://eur-lex.europa.eu/legal-content/EN/TXT/?uri=CELEX%3A12007L%2FTXT

European Council (2003), *European Security Strategy: A Secure Europe in a Better World,* Brussels, 12 December 2003 [Not Published in the Official Journal], https://www.consilium.europa.eu/media/30823/qc7809568enc.pdf

European External Action Service (2016), *The Global Strategy for the Foreign and Security Policy of the European Union,* https://eeas.europa.eu/sites/eeas/files/eugs_review_web_0.pdf

European Parliament and of the Council of the European Union (2016), *Directive 2016/1148 of the European Parliament and of the Council Concerning Measures for a High Common Level of Security of Network and Information Systems Across the Union,* EUR-Lex, Official Journal, https://eur-lex.europa.eu/legal-content/EN/TXT/PDF/?uri=CELEX:32016L1148&from=EN

THIRD CHAPTER (Croatia)

Croatian Parliament (2018), *Law on the Cyber Security of Key Service Operators and Digital Services Providers.* Official Gazette, No. 64/2018, https://narodne-novine.nn.hr/clanci/sluzbeni/2018_07_64_1305.html (Croat.), Accessed: 26 August 2019

Croatian Parliament (2017a), *National Security Strategy.* Official Gazette, No. 73/2017, https://www.soa.hr/files/file/National-Security-Strategy-2017.pdf, Accessed: 27 August 2019

Croatian Parliament (2017b), *Homeland Security Law.* Official Gazette, No. 108/2017, https://www.zakon.hr/z/955/Zakon-o-sustavu-domovinske-sigurnosti, (Croat.), Accessed: 27 August 2019

Croatian Parliament (2013), *Critical Infrastructure Law.* Official Gazette, No. 56/2013, http://www.zakon.hr/z/591/Zakon-o-kriti%C4%8Dnim-infrastrukturama (Croat.), Accessed: 25 August 2019

Croatian Parliament (2012), *Public-Private Partnership Law.* Official Gazette, No. 78/12, 152/14, 114/18, https://www.zakon.hr/z/198/Zakon-o-javno-privatnom-partnerstvu, (Croat.), Accessed: 27 August 2019

Croatian Parliament (2010), *Private Protection Law.* Official Gazette, No. 68/03, 31/10, 139/10, https://www.zakon.hr/z/291/Zakon-o-privatnoj-zaštiti, (Croat.), Accessed: 27 August 2019

European Parliament and of the Council of the European Union (2016), *Directive 2016/1148 of the European Parliament and of the Council Concerning Measures for a High Common Level of Security of Network and Information Systems Across the Union,* EUR-Lex, Official Journal, https://eur-lex.europa.eu/legal-content/EN/TXT/PDF/?uri=CELEX:32016L1148&from=EN, Accessed: 29 August 2019

Government of the Republic of Croatia (2015a), *National Strategy for the Prevention and Countering of Terrorism*. Official Gazette, No. 108/2015, https://narodne-novine.nn.hr/clanci/sluzbeni/2015_10_108_2105.html, (Croat.), Accessed: 12 September 2019

Government of the Republic of Croatia (2015b), *Croatian National Cyber Security Strategy*. Official Gazette, No. 108/2015, https://narodne-novine.nn.hr/clanci/sluzbeni/2015_10_108_2106.html, (Croat.), Accessed: 25 August 2019

Government of the Republic of Croatia (2013a), *Decision on Designation the Sectors from Which the Central State Administrative Bodies Identify National Critical Infrastructure and Lists of the Order of the Sectors of Critical Infrastructures*. Official Gazette, No. 108/2013, http://narodne-novine.nn.hr/clanci/sluzbeni/2013_08_108_2411.html (Croat.), Accessed: 25 August 2019

Government of the Republic of Croatia (2013b), *Risk Assessment for Republic of Croatia from Natural and Technical-Technological Disasters and Major Accidents*, https://civilna-zastita.gov.hr/UserDocsImages/DOKUMENTI_PREBACIVANJE/PLANSKI%20DOKUMENTI%20I%20UREDBE/Procjena%20rizika%20od%20katastrofa%20za%20RH.pdf /, (Croat.), Accessed: 29 August 2019

Government of the Republic of Croatia (2010), *Protection and Rescue Plan for the Republic of Croatia*. Official Gazette, No. 96/2010, https://narodne-novine.nn.hr/clanci/sluzbeni/2010_08_96_2707.html, (Croat.), Accessed: 29 August 2019

Government of the Republic of Croatia (2009), *National Strategy and Action Plan for the Non-Proliferation of Weapons of Mass Destruction*. Government Session Held on 10 September 2009 (Not Published in Official Gazette), https://vlada.gov.hr/UserDocsImages//2016/Sjednice/Arhiva//71.%20-%206.pdf, (Croat.), Accessed: 12 September 2019

Government of the Republic of Croatia (2008), *National Strategy for the Prevention and Countering of Terrorism*. Official Gazette, No. 139/2008, http://zakon.poslovna.hr/public/nacionalna-strategija-za-prevenciju-i-suzbijanje-terorizma/466704/zakoni.aspx, (Croat.), Accessed: 29 August 2019

Matika, Dario; Poljanec Borić, Saša (Eds.) (2009), *Critical Infrastructure in Croatia*. Zagreb: Institute for Research and Development of Defense Systems of the Ministry of Defense and Ivo Pilar Institute of Social Sciences, https://www.pilar.hr/wp-content/images/stories/dokumenti/kriticna_infrastruktura/eski_knjiga_cd_003.pdf (Croat.)

Mikac, Robert; Cesarec, Ivana; Larkin, Richard (2018), *Critical Infrastructure Systems: The Platforms for a Successful Development of Nations Security*, Zagreb: Jesenski and Turk (Croat.)

Ministry of Defense (1999), *Rules on Criteria for Designating and Protecting Objects Particularly Significant for the Defense of the Country*. Official Gazette, No. 120/1999, https://narodne-novine.nn.hr/clanci/sluzbeni/1999_11_120_1936.html (Croat.), Accessed: 26 August 2019

National Protection and Rescue Directorate (2016), *Rules on Methodology for Drafting Business Risk Analysis of Critical Infrastructure*. Official Gazette, No. 47/2016, http://www.poslovni-savjetnik.com/propisi/pravilnik-o-metodologiji-za-izradu-analize-rizika-poslovanja-kriticnih-infrastruktura-vazeci (Croat.), Accessed: 25 August 2019

National Protection and Rescue Directorate (2013), *Rules on the Methodology for Drafting Business Risk Analysis of Critical Infrastructure*. Official Gazette, No. 128/2013, http://narodne-novine.nn.hr/clanci/sluzbeni/2013_10_128_2792.html (Croat.), Accessed: 25 August 2019

FOURTH CHAPTER (Turkey)

AFAD (2014), *2014–2023 CIP Roadmap Document*. Ankara: AFAD

AFAD (2019), *About Us*. Retrieved from Disaster and Emergency Management Authority (AFAD), https://www.afad.gov.tr/en/2572/About-Us

Afyonluoğlu, Mustafa (2017), *Türkiye Kritik Altyapılarda Siber Güvenlik Açısından Neler Yapmal*, https://afyonluoglu.org/PublicWebFiles/presentations/Bilisim2017-Kritik%20Altyap%C4%B1lar-Mustafa%20Afyonluoglu.pdf

Ak, Tarik (2019), İç Güvenlik Yönetimi Açısından Kritik Altyapıların Korunması. *ASSAM Uluslararası Hakemli Dergi*, 42–51

Bıçakcı, S. Salih; Ergun, Doruk; Çelikpala, Mitat (2015), The Cyber Security Scene in Turkey. In Bıçakcı, Salih, et al. (Eds.), *A Primer on Cyber Security in Turkey and the Case of Nuclear Power* (pp. 22–51). İstanbul: EDAM

BT Haber (2020), *Türkiye'nin Siber Güvenlik Üssü Cumhurbaşkanımız Recep Tayyip Erdoğan'ın Teşrifleri İle Açılıyor*, https://www.bthaber.com/turkiyenin-siber-guvenlik-ussu-cumhurbaskanimiz-recep-tayyip-erdoganin-tesrifleri-ile-aciliyor/

Çelikpala, Mitat (2013), Terrorist Threats to Energy Critical Infrastructure Technology and Protection Systems. In Aydoğdu, F. U. (Ed.), *Technological Dimensions of Defence Against Terrorism* (pp. 81–90). IOS Press. doi:10.3233/978-1-61499-317-9-81

Çelikpala, Mitat (2009), Protecting the Key National Utilities and Energy Infrastructure. In Ker-Lindsay, J., Cameron, A. (Eds.), *Combating International Terrorism Turkey's Added Value* (pp. 16–19). London: Royal United Services Institute

CISA (2019), *A Guide to Critical Infrastructure Security and Resilience*. Washington, DC: Cybersecurity Infrastructure Security Agency

Council of the European Union (2020), *Turkey*, https://www.consilium.europa.eu/policies/enlargement/turkey.aspx/

CSIS (2013), *Public-Private Partnerships for Critical Infrastructure Protection*. Washington, DC: Center for Strategic International Studies (CSIS), https://www.csis.org/analysis/public-private-partnerships-critical-infrastructure-protection-0

Cybermag (2018), *EPDK Bilişim Hizmetleri Grup Başkanı Mehmet Yılmazer ile Söyleşi*, https://www.cybermagonline.com/epdk-bilisim-hizmetleri-grup-baskani-mehmet-yilmazer-ile-soylesi

Digital Transformation Office of the Presidency of the Republic of Turkey (2020a), *Bilgi ve İletişim Güvenliği Rehberi*. Ankara: Digital Transformation Office of the Presidency of the Republic of Turkey, https://cbddo.gov.tr/bgrehber

Digital Transformation Office of the Presidency of the Republic of Turkey (2020b), *Siber Güvenlik Dairesi Başkanlığı*, https://cbddo.gov.tr/hizmet-birimlerimiz/siber-guvenlik-dairesi-baskanligi/

Dunn-Cavelty, Myriam; Suter, Manuel (2009), Public–Private Partnerships Are No Silver Bullet: An Expanded Governance Model for Critical Infrastructure Protection. *International Journal of Critical Infrastructure Protection*, 179–187

ENISA (2017), *Public Private Partnerships (PPP) Cooperative Models*. Athens, Greece

European Commission (2020), *EU Grants €38 Million for Protection of Critical Infrastructure Against Cyber Threats*, https://ec.europa.eu/programmes/horizon2020/en/news/eu-grants-%E2%82%AC38-million-protection-critical-infrastructure-against-cyber-threats#:~:text=The%20Commission%20announced%20today%20that,making%20cities%20smarter%20and%20safer

European Commission Expert Group on Electricity Interconnection Targets (2019), *Electricity Interconnections with Neighbouring Countries*, https://op.europa.eu/en/publication-detail/-/publication/785f224b-93cd-11e9-9369–01aa75ed71a1, Accessed: 20 May 2020

Fleming, Matthew; Goldstein, Eric; Roman, John (2014), *Evaluating the Impact of Cybersecurity Information Sharing on Cyber Incidents and Their Consequences*. USA: Homeland Security Studies and Analysis Institute.

Gürtaş, Muhammed Ali (2019), *Turkey: 1M+ Health Care Professionals in 2018*, https://www.aa.com.tr/en/health/turkey-1m-health-care-professionals-in-2018/1598912

Hürriyet (2014), *Özel sektör elektrik üretiminde tavan yaptı*, https://www.hurriyet.com.tr/ekonomi/ozel-sektor-elektrik-uretiminde-tavan-yapti-27258584

IBRD & World Bank (2018), *High-Level Critical Infrastructure Risk Assessment (CIRA) in the Region of Çukurova (Turkey)*. Washington, DC: International Bank for Reconstruction and Development

ITProportal (2015), *Turkish Blackout Sparks Fears of Cyber Attack on the West*, https://www.itproportal.com/2015/05/19/turkish-blackout-sparks-fears-cyber-attack-on-west/

Karabacak, Bilge; Ozkan Yildirim, Sevgi; Baykal, Nazife (2016), Regulatory Approaches for Cyber Security of Critical Infrastructures: The Case of Turkey. *Computer Law & Security Review*, 526–539

Kasapoğlu, Can (2015), Turkey's Future Cyber Defense Landscape. In Bıçakcı, S., et al. (Eds.), *A Primer on Cyber Security in Turkey and the Case of Nuclear Power* (pp. 2–20). İstanbul: EDAM

KMO Information Systems (2020), *TA for Achieving the Harmony with EU Regarding the Implementation of the NIS Directive*, http://www.kmo.com.tr/en/39062/TA-for-Achieving-the-Harmony-with-EU-regarding-the-Implementation-of-the-NIS-Directive

Milliyet (2018), *Siber Güvenlik Ilkokula Giriyor*, https://www.milliyet.com.tr/gundem/siber-guvenlik-ilkokula-giriyor-2778996

Ministry of Transport and Infrastructure (2016), *2016–2019 National Cyber Security Strategy, Turkey*

Ministry of Transport and Infrastructure (2014), *Sektörel SOME Kurulum ve Yönetim Rehberi*. doi:https://www.uab.gov.tr/uploads/pages/siber-guvenlik/sektorel-some-reh.pdf

OECD (2018), *Joint OECD – EU JRC Workshop: System Thinking for Critical Infrastructure Resilience and Security, France*

OECD (2017), *Turkey: Critical Risks at a Glance*, http://www.oecd.org/gov/risk/risk-governance-country-factsheet-turkey.pdf

Official Gazette of Turkey (2019), *Bilgi ve İletişim Güvenliği Tedbirleri*, https://www.resmigazete.gov.tr/eskiler/2019/07/20190706-10.pdf

Official Gazette of Turkey (2017), *Enerji Sektöründe Kullanılan Endüstriyel Kontrol Sistemlerinde Bilişim Güvenliği Yönetmeliği*

Official Gazette of Turkey (2013), *Siber Olaylara Müdahale Ekiplerinin Kuruluş, Görev Ve Çalişmalarina Dair Usul Ve Esaslar Hakkinda Tebliğ*, https://www.mevzuat.gov.tr/mevzuat?MevzuatNo=19004&MevzuatTur=9&MevzuatTertip=5

Oliver, David, A.; Haney, Michael (2017), *Curriculum Development for Teaching Critical Infrastructure Protection 21st Colloquium for Information Systems Security Education*. USA: Idaho National Laboratory

OSCE (2013), *Good Practices Guide on Non-Nuclear Critical Energy Infrastructure Protection (NNCEIP) from Terrorist Attacks Focusing on Threats Emanating from Cyberspace, Austria*
Pollet, J. Jonathan; Cummins, Joe (2009), All Hazards Approach for Assessing Readiness of Critical Infrastructure. In *IEEE Conference on Technologies for Homeland Security* (pp. 366–372). Boston, MA: IEEE. doi:10.1109/THS.2009.5168060
Presidency of the Republic of Turkey Investment Office (2019), *Investing in Infrastructure & Public Private Partnership (PPP) in Turkey*, https://www.invest.gov.tr/en/library/publications/lists/investpublications/infrastructure-industry.pdf
Şeker, Ensar; Tolga, İhsan Burak (2018), *National Cyber Security Organization: Turkey*
STM (2019), *About*, https://ctf.stm.com.tr/about
TÜBİTAK (2017), *Kritik Bilgi Sistem Altyapıları için Asgari Güvenlik Önlemleri Dokümanı*, https://hgm.uab.gov.tr/uploads/pages/siber-guvenlik/kritik-bilgi-sistem-altyapilari-i-c-in-asgari-gu-venlik-o-nlemleri-6445b90e-b2ad-4e5e-9c13–6ae19ba10e37.pdf
Türkiye Siber Güvenlik Kümelenmesi (2020), https://www.siberkume.org.tr/Index
United Nations CTED and UNOCT (2018), *The Protection of Critical Infrastructures Against Terrorist Attacks: Compendium of Good Practices*, https://www.un.org/sc/ctc/wp-content/uploads/2018/06/Compendium-CIP-final-version-120618_new_fonts_18_june_2018_optimized.pdf
World Bank (2018), *Supporting Critical Infrastructure Resilience and Disaster-Risk Awareness in Turkey Project Information Document*, http://documents.worldbank.org/curated/en/461891533554597941/pdf/Project-Information-Document-PID-Supporting-Critical-Infrastructure-Resilience-and-Disaster-Risk-Awareness-in-Turkey-P168167.pdf
Zio, E. Enrico (2018), *A Systemic View to Critical Infrastructure Resilience*, http://www.oecd.org/gov/risk/enrico-zio-session-one.pdf

FIFTH CHAPTER (Bosnia and Herzegovina)

Blueprint Energy Solutions GmbH (2019), *Final Report – Study on Cyber Security in the Energy Sector of the Energy Community*, https://www.energy-community.org/dam/jcr:db8e479d-b423-40c9-9ff9-998c7d9045ef/Blueprint_cyber_122019.pdf
Cikotić, S. Selmo; Smajić, Mirza; Delić, Haris; Subašić, Nihad (2018), *National Security and Private Protection*. Sarajevo: Faculty of Political Science
Council of the European Union (2008), *Council Directive 2008/114/EC of 8 December 2008 on the Identification and Designation of European Critical Infrastructures and the Assessment of the Need to Improve Their Protection*. Brussels, 2008/114/EC, EUR-Lex, Official Journal, https://eur-lex.europa.eu/LexUriServ/LexUriServ.do?uri=OJ:L:2008:345:0075:0082:EN:PDF
Council of Ministers of Bosnia and Herzegovina (2017), *Information Security Managment Polices in the Institutions of Bosnia and Herzegowina 2017–2022*. Official Gazette, No. 38/17, http://www.sluzbenilist.ba/page/akt/bM0k8gNBNCU=
Council of Ministers of Bosnia and Herzegovina (2011), *Risk Assessment for Bosnia and Herzegovina Regarding Natural and Other Disasters*. Official Gazette, No. 54/10, http://www.msb.gov.ba/PDF/PROCJENA_UGRO%C5%BDENOSTI_BIH_07102013.pdf
Council of Ministers of Bosnia and Herzegovina (2009), *Methodology for the Preparation of Risk Assessment of Bosnia and Herzegovina Regarding Natural and Other Disasters*. Official Gazette, No. 86/09, http://msb.gov.ba/docs/bosanski.doc

European Parliament (2020), *The Western Balkans, Fact Sheets*, https://www.europarl.europa.eu/factsheets/en/sheet/168/the-western-balkans

Italian Agency for Development Cooperation; United Nations Development Programme (2018), *Disasters Response Capacities Assessment and the Roadmap for Bosnia and Herzegovina*, http://fucz.gov.ba/wp-content/uploads/2019/02/procjena-kapaciteta-za-odgovor-na-katastrofe-i-mapa-puta-za-bih.pdf

Lisica, Darvin (2011), *Security Risks and Fundamental Social Values in Bosnia and Herzegovina*. Sarajevo: Faculty of Political Science

Mikac, Robert; Cesarec, Ivana; Larkin, Richard (2018), *Critical Infrastructure Systems: The Platforms for a Successful Development of Nations Security*. Zagreb: Jesenski and Turk (Croat.)

Ministry of Security of Bosnia and Herzegovina (2015), *Strategy of Bosnia and Herzegovina for Prevention and Fight Against Terrorism 2015–2020*, http://msb.gov.ba/dokumenti/strateski/default.aspx?id=12786&langTag=bs-BA

Ministry of Security of Bosnia and Herzegovina (2010), *Strategy of Bosnia and Herzegovina for Prevention and Fight Against Terrorism 2010–2013*, http://www.psa.gov.ba/Popisi_i_dokumenti/Propisi/Strategije/Strategija%20BiH%20za%20prevenciju%20i%20borbu%20protiv%20terorizma.pdf

National Assembly of the Republic Srpska (2019), *Law on Security of Critical Infrastructures in Republic Srpska*. Official Gazette, No. 58/19, https://www.narodnaskupstinars.net/?q=la/akti/usvojeni-zakoni/zakon-o-bezbjednosti-kriti%C4%8Dnih-infrastruktura-u-republici-srpskoj

OSCE in BiH (2018), *National Table Top Exercise in Bosnia and Herzegovina on Protecting Critical Energy Infrastructure from Cyber-Related Terrorist Attacks*, https://polis.osce.org/national-table-top-exercise-bosnia-and-herzegovina-protecting-critical-energy-infrastructure

Presidency of Bosnia and Herzegovina (2006), *Security Policy of Bosnia and Herzegovina*, http://www.mod.gov.ba/slike2014/02.28.20_1_Sigurnosna%20politika%20BiH.pdf

Sarajevogas Company (2020), *About the Company*, https://www.sarajevogas.ba/distribucijaprirodnoggasa

Smajić, Mirza (2018), Multiculturalism in Bosnia and Herzegowina, Book of Proceedings Multiculturalism in Public Policies. In Bašić, G., Žagar, M., Tatalović, S. (Eds.), *Academic Network for Cooperation in South-East Europe* (pp.128–149). Belgrade: Institute of Social Sciences Belgrade, Institute for Ethnic Studies Ljubljana, Centre for International and Security Studies, Faculity of Political Scineces, Zagreb, https://idn.org.rs/biblioteka/Multiculturalism_in_public_policies_sajt.pdf

Smajić, Mirza (2017), *Public-Private Partnership and Critical Infrastructure in the Field of Security in Bosnia and Herzegovina*. Zagreb: International Conference Private (Military) and Security Companies in Critical Infrastructure Protection, https://www.racviac.org/news/news17.html

SIXTH CHAPTER (Serbia)

Center for Euro-Atlantic Studies, OSCE Mission in Serbia (2016), *Guide to Information Security in the Republic of Serbia*, https://www.ceasserbia.org/images/2016/09/Vodic_kroz_informacionu_bezbednost_SRB_Web_1.pdf, Accessed: 7 November 2020.

European Commission (2013), *Commission Staff Working Document on a New Approach to the European Programme for Critical Infrastructure Protection Making European Critical Infrastructures More Secure*, https://ec.europa.eu/energy/sites/ener/files/documents/20130828_epcip_commission_staff_working_document.pdf

European Commission (2010), *Communication from the Commission to the European Parliament and the Council - The EU Internal Security Strategy in Action: Five Steps Towards a More Secure Europe*, EUR-Lex, Official Journal, https://eur-lex.europa.eu/legal-content/EN/TXT/?uri=celex%3A52010DC0673

European Commission (2006), *Communication from the Commission on a European Programme for Critical Infrastructure Protection,* https://eur-lex.europa.eu/legal-content/EN/TXT/HTML/?uri=CELEX:52006DC0786&from=EN

European Council (2009), Conclusions of the European Council of 10/11 December 2009 on 'The Stockholm Programme – An Open and Secure Europe Serving and Protecting Citizens (2010–2014); 17024/09, Accessed: 07 January 2016

European Council, Council Directive 2008/114/EC of December 8.2008, on the Identification and Designation of European Critical Infrastructures and the Assessment of the Need to Improve Their Protection (SL L 345/75, 23.12.2008.).

Forrer, John; Kee, James Edwin; Newcomer, Kathryn E.; Boyer, Eric (2014), Public Private Partnerships and the Public Accountability Question. In Boyer, E., et al. (Eds., p. 4), https://sites.duke.edu/niou/files/2011/05/Forrer-Lee-Newcomer-and-Boyer-Public-Private-Partnership-and-the-Public-Accountability-Question.pdf, Accessed: 7 November 2020

Government of Republic of Serbia (2019a), *Decision on Identification of Products and Services of Special Importance for the National Defence of the Republic of Serbia*. Official Gazette, No. 58/2008, 26/2019, http://www.pravno-informacioni-sistem.rs/SlGlasnikPortal/eli/rep/sgrs/vlada/odluka/2008/58/2/reg, Accessed: 20 July 2020.

Government of Republic of Serbia (2019b), *Guidelines on Methodology for the Development and Content of Disaster Risk Assessment and Protection and Rescue Plans*. Official Gazette, No. 80/2019, https://www.paragraf.rs/propisi/uputstvo-metodologiji-izradu-procene-ugrozenosti-elementarnih-nepogoda.html, Accessed: 20 July 2020.

Government of Republic of Serbia (2015), *Decision on Identification of Large Technical Systems Important for National Defence*. Official Gazette, No. 41/2014, 35/2015, https://www.pravno-informacioni-sistem.rs/SlGlasnikPortal/eli/rep/sgrs/vlada/odluka/2014/41/1/reg, Accessed: 20 July 2020.

Government of Republic of Serbia (2012), *Guideline on Methodology for the Risk Assessment and the Protection and Rescue Plans in a State of Emergency*. Official Gazette, No. 96/2012, https://www.kopering.rs/images/propisi/uputstvo_o_metodologiji_za_izradu_procene_ugrozeosti_i_planova_zastite_i_spasavanja_u_vanrednim_situacijama.pdf, Accessed: 20 July 2020.

Government of Republic of Serbia (2011), *Regulation on the Content and Methodology for the Development of Protection and Rescue Plans in Emergency Situations*. Official Gazette, No. 8/2011, https://pinbezbednost.rs/wp-content/uploads/2019/03/Uredba-o-sadr%C5%BEaju-i-na%C4%8Dinu-izrade-planova-za%C5%A1tite-i-spasavanja-u-vanrednim-situacijama.pdf, Accessed: 20 July 2020

Government of Republic of Serbia (2008), *Decision on Facilities of Particular Importance for National Defence*. Official Gazette, No. 112/2008, https://www.pravno-informacioni-sistem.rs/SlGlasnikPortal/eli/rep/sgrs/vlada/odluka/2008/112/1/reg, Accessed: 20 July 2020

Institut for Standardization of Serbia, *SRPS A.L2.003:2009, Societal Security – Private Security Services – Requirements and Guidelines for Conformity Assessment.* Official Gazette, No. 07/2009.

International Organization for Standardization, *ISO 31000 Risk Management – Principles and Guidelines on Implementation.*

International Organization for Standardization, *ISO/CD 3 22397 Societal Security – Guidelines for Establishing Partnering Arrangements.*

International Organization for Standardization, *ISO/IEC 31010 Risk Management – Risk Assessment Techniques.*

International Organization for Standardization, *ISO/TC 223 N 337 Societal Security - Public Private Partnership – Guidelines for Establishing Partnership Agreements Among Organizations.*

Keković, Zoran; Ninković, Vladimir (2020), Towards a Conceptualisation of Resilience in Security Studies. *Serbian Political Thought*, 67, pp. 153–175

National Assembly of Republic of Serbia (2020), *Law on Planning and Construction.* Official Gazette, 24/2011, 121/2012, 42/2013, 132/2014, 145/2014, 83/2018, 31/2019, 37/2019, 9/2020, https://www.paragraf.rs/propisi/zakon_o_planiranju_i_izgradnji.html, Accessed: 20 July 2020

National Assembly of Republic of Serbia (2019), *National Security Strategy.* Official Gazette, No. 139/2008, https://www.pravno-informacioni sistem.rs/SlGlasnikPortal/eli/rep/sgrs/skupstina/strategija/2019/94/2, Accessed: 20 July 2020

National Assembly of Republic of Serbia (2018a), *Law on Defence.* Official Gazette, No. 116/2007, 88/2009, 88/2009, 104/2009, 10 / 2015, 36/2018, https://www.paragraf.rs/propisi/zakon_o_odbrani.html, Accessed: 20 July 2020.

National Assembly of Republic of Serbia (2018b), *The Law on Disaster Risk Reduction and Emergency Management.* Official Gazette, No. 87/2018, https://www.paragraf.rs/propisi/uputstvo-metodologiji-izradu-procene-ugrozenosti-elementarnih-nepogoda.html, Accessed: 20 July 2020.

National Assembly of Republic of Serbia (2018c), *Law on Environmental Protection.* Official Gazette, No. 135/2004, 36/2009, 36 / 2009, 72/2009, 43/2011, 14/2016, 76/2018, 95/2018, https://www.paragraf.rs/propisi/zakon_o_zastiti_zivotne_sredine.html, Accessed: 20 July 2020.

National Assembly of Republic of Serbia (2018d), *Law on Water.* Official Gazette, No. 30/2010, 93/2012, 101/2016, 95/2018 95/2018, https://www.paragraf.rs/propisi/zakon_o_vodama.html, Accessed: 20 July 2020

National Assembly of Republic of Serbia (2018e), *Law on Critical Infrastructure.* Official Gazette, No. 87/2018, https://www.paragraf.rs/propisi/zakon-o-kriticnoj-infrastrukturi.html, Accessed: 20 July 2020.

National Assembly of Republic of Serbia (2016a), *Law on Information Security.* Official Gazette, No. 6/2016, 94/2017 77/2019, https://www.paragraf.rs/propisi/zakon_o_informacionoj_bezbednosti.html, Accessed: 7 November 2020.

National Assembly of Republic of Serbia (2016b), *Law on Public-Private Partnerships and Concessions.* Official Gazette, No. 88/2011, 15/2016, 104/2016, http://www.paragraf.rs/propisi/zakon_o_javno_privatnom_partnerstvu_i_koncesijama.html, Accessed: 2 January 2016.

National Assembly of Republic of Serbia (2013), *Law on Private Security.* Official Gazette, No. 104/2013, 42/2015, 87/2018, https://www.paragraf.rs/propisi/zakon_o_privatnom_obezbedjenju.html, Accessed: 20 July 2020

National Assembly of Republic of Serbia (2011), *National Strategy of Protection and Rescue in Emergency Situations*. Official Gazette, No. 86/2011, http://arhiva.mup.gov.rs/cms_lat/sadrzaj.nsf/Nacionalna_strategija_zastite_i_spasavanja_u_vanrednim_situacijama_lat.pdf, Accessed: 20 July 2020

National Assembly of Republic of Serbia (2009), *Data Secrecy Law*. Official Gazette, No. 104/2009, https://www.paragraf.rs/propisi/zakon_o_tajnosti_podataka.html, Accessed: 20 July 2020.

SEVENTH CHAPTER (Montenegro)

Elektroprivreda Crne Gore, https://www.epcg.com/

EPCG (2016), *Elektroprivreda - List EPCG*, No. 365

Government of Montenegro (2019a), *Law on Determination and Protection of Critical Infrastructure of Montenegro*. Official Gazette, No. 72/2019

Government of Montenegro (2019b), *Subcommittee on Transport, Environment, Energy and Regional Development Between Montenegro and the European Commission*, http://www.gov.me/naslovna/vijesti-iz-ministarstava/202400/Pododbor-za-saobracaj-zivotnu-sredinu-energetiku-i-regionalni-razvoj-izmedu-Crne-Gore-i-Evropske-komisije.html

Government of Montenegro (2018a), *National Security Strategy of Montenegro with Action Plan*, https://kei.gov.me/ResourceManager/FileDownload.aspx?rid=161427&rType=2&file=Strategy%20of%20National%20Security%20of%20Montenegro".pdf

Government of Montenegro (2018b), *Law on Protection of Persons and Objects*. Official Gazette, No. 43/18

Government of Montenegro (2017a), *Cyber Security Strategy of Montenegro 2018–2021*, http://www.mju.gov.me/ResourceManager/FileDownload.aspx?rid=305198&rType=2&file=Cyber%20Security%20Strategy%20of%20Montenegro%202018-2021%20eng.pdf

Government of Montenegro (2017b), *Strategy for disaster risk reduction (2018-2023)*, https://wapi.gov.me/download/a0608df5-ff2a-465b-a733-1830a60e35ff?version=1.0

Government of Montenegro (2016a), *Law on Protection and Rescue*. Official Gazette, No. 013/07, 005/08, 086/09, 032/11, 54/16

Government of Montenegro (2016b), *Law on Cyber Security*. Official Gazette, No. 014/10, 040/16

Government of Montenegro (2009), *Law on State Property*. Official Gazette, No. 21/09

Mikac, Robert; Cesarec, Ivana; Larkin, Richard (2018), *Critical Infrastructure Systems: The Platforms for a Successful Development of Nations Security*. Zagreb: Jesenski and Turk (Croat.).

Ministry of Information Society and Telecommunications of Montenegro (2014), *Critical IT Infrastructure Selection Methodology*

Mitrevska, Marina; Mileski, Toni; Mikac, Robert (2019), *Critical Infrastructure: Concept and Security Challenges*. Skoplje: Friedrich Ebert Foundation

Monteput (2020), http://monteput.me/

Urban Protection Training Center (UPTC) (2019), https://www.uptc.me/

EIGHT CHAPTER (North Macedonia)

Alcheski, G. J. (2016), *Implementacija na sovremenite bezbednosni sistemi i procedure vo razvojot na obezbeduvanjeto na objektite od vitalen interes za Republika Makedonija (so osvrt na aerodromskata bezbednost)* (pp. 213–214). Skopje: Filozofski fakultet

Government of North Macedonia (2020), *Defence Law*. Official Gazette, No. 42/2020
Government of North Macedonia (2018), *Protection and Rescue Law*. Official Gazette, No.83/2018
Government of North Macedonia (2015a), *Law on Internal Affairs*. Official Gazette, No. 33/2015
Government of North Macedonia (2015b), *Aviation Law*. Official Gazette, No. 63/2015
Government of North Macedonia (2013), *Law on Private Security*. Official Gazette, No.106/2013
Government of North Macedonia (2012), *Law on Private Security*. Official Gazette, No. 166/2012
Government of North Macedonia (2005), *Crisis Management Law*. Official, No. 29/2005
Ministry of Information Society and Administration (2018), *National Cyber Security Strategy of the Republic of Macedonia (2018–2022)*.
Mitrevska, Marina; Mikac, Robert (2017), *Handbook on Critical Infrastructure Protection*. Skopje: Chamber of Republic of Macedonia for Private Security
Mitrevska, Marina; Mileski, Toni; Mikac, Robert (2019), *Critical Infrastructure: Concept and Security Challenges*. Skoplje: Friedrich Ebert Foundation

NINTH CHAPTER (Albania)

Albanian Assembly (2019), *Law No. 45/2019 "On Civil Protection"*. Official Gazette, No. 113, https://qbz.gov.al/eli/fz/2019/113/837b2b34-2833-4add-8635-994462b57cc5
Albanian Assembly (2018), *Law No. 8/2018 "On the Ratification of the Treaty Establishing the Transport Community"*. Official Gazette, No. 36, https://qbz.gov.al/eli/fz/2018/36/ff5afa83-2a15-46ea-a0b0-5713f8a10f05
Albanian Assembly (2017), *Law No. 2/2017 "On Cyber Security"*. Official Gazette, No. 22, https://qbz.gov.al/eli/fz/2017/22/f1d86587-1880-4d99-970c-f4a52cf64ce3
Albanian Assembly (2015a), *Law No. 152/2015, "On the Fire Protection and Rescue Service"*. Official Gazette No. 240, Dated 07.01.2016, https://qbz.gov.al/eli/fz/2015/240/03ba472b-3a09-4363-a68b-01f4ccff976e
Albanian Assembly (2015b), *Law No. 43/2015 "On the Electricity Sector"*. Official Gazette, No. 87, https://qbz.gov.al/eli/fz/2015/87/9c61d502-229c-4370-82d0-1fecd91a185b
Albanian Assembly (2015c), *Law No. 102/2015 "On the Natural Gas Sector"*. Official Gazette, No. 178, https://qbz.gov.al/eli/fz/2015/178/f80284bc-fdb3-4927-8429-4f240a208d3b
Albanian Assembly (2015d), Law No. 77/2015 *"On Some Changes and Amendments on the Law No. 125/2013 "On Concessions and Public-Private Partnership"*. Official Gazette, No. 132, https://qbz.gov.al/eli/fz/2015/132/b6e17bb8-23ca-4d03-8d98-18b5403ec212
Albanian Assembly (2014a), *Law No. 103/2014 "On the Approval of the National Security Strategy of the Republic of Albania"*. Official Gazette, No. 137, https://qbz.gov.al/eli/fz/2014/137/b2c719c8-a0df-46df-ac10-c3dda9bc049b
Albanian Assembly (2014b), *Law No. 107/2014, "On Territorial Planning and Development"*. Official Gazette, No. 137, https://qbz.gov.al/eli/fz/2014/137/b2c719c8-a0df-46df-ac10-c3dda9bc049b
Albanian Assembly (2013a), *Law No. 157/2013 "On Measures Against the Financing of Terrorism"*. Official Gazette, No. 172, https://qbz.gov.al/eli/fz/2013/172/cc8d9504-0011-4c47-a166-6e583dc4e93a

Albanian Assembly (2013b), *Law No. 125/2013 "On Concessions and Public-Private Partnership"*. Official Gazette, No. 76, https://qbz.gov.al/eli/fz/2013/76/a83ff478-3303-463d-8b38-ae716e6b4bee

Albanian Assembly (2012), *Law No. 102/2012 on Some Changes and Additions to Law No. 9918, "On Electronic Communications in the Republic of Albania"*. Official Gazette, No. 145, https://qbz.gov.al/eli/fz/2012/145/02169e53-3187-4dbe-b98a-1ac03efe752b

Albanian Assembly (2008), *Law No. 9918 "On Electronic Communications in the Republic of Albania"*. Official Gazette, No. 84, https://qbz.gov.al/eli/fz/2008/84/be85e8c5-2641-4238-8a71-26bdb71b5b00

Albanian Assembly (2006a), *Law No. 9501, "On the Ratification of the Treaty Establishing the Energy Community"*. Official Gazette, No. 38, https://qbz.gov.al/eli/fz/2006/38/eeae59c9-3087-4f55-8fe7-9576521e183d

Albanian Assembly (2006b), *Law No. 9590 "On the Ratification of the Stabilization-Association Agreement Between the Republic of Albania and the European Communities and Their Member States"*. Official Gazette, No. 87, https://qbz.gov.al/eli/fz/2006/87/3e565b9e-375a-4884-9897-6e75854aa0c4

Blueprint Energy Solutions GmbH (2019), *Study on Cyber Security in the Energy Sector of the Energy Community, Final Report December 2019*, https://www.energy-community.org/dam/jcr:db8e479d-b423-40c9-9ff9-998c7d9045ef/Blueprint_cyber_122019.pdf

Council of the European Union (2008), *Council Directive 2008/114/EC of 8 December 2008 on the Identification and Designation of European Critical Infrastructures and the Assessment of the Need to Improve Their Protection,* EUR-Lex, Official Journal, https://eur-lex.europa.eu/LexUriServ/LexUriServ.do?uri=OJ:L:2008:345:0075:0082:EN:PDF

Council of Ministers (2020), *Decision No. 553 "On the Approval of the List of Critical Information Infrastructures and the List of Important Information Infrastructures"*. Official Gazette, No. 134, https://qbz.gov.al/eli/fz/2020/134/cb5a5a72-3cb8-49f5-aaee-219948f63abc

Council of Ministers (2019a), *Decision No. 747 "On the Organization and Functioning of the National Agency of Civil Protection"*. Official Gazette, No. 159, https://qbz.gov.al/eli/fz/2019/159/e3c16f4c-2af2-47ac-b55b-9073c24561f9

Council of Ministers (2019b), *Decision No. 520 "On the Approval of the Regulation for the Fire Protection and Rescue Service"*. Official Gazette, No. 112, https://qbz.gov.al/eli/fz/2019/112/201d59f2-30ed-433f-a38c-a9fae241b82b

Council of Ministers (2018a), *Decision No. 480, Dated 31 July 2018, "On the Approval of the National Energy Strategy for the Period 2018–2030"*. Official Gazette, No. 119, Dated 9.08.2018, https://qbz.gov.al/eli/fz/2018/119/43edafea-5386-450b-83c2-f434d0df229c

Council of Ministers (2018b), *Decision No. 87 "On the Approval of the Development Plan of the Natural Gas Sector in Albania and the Identification of Priority Projects"*. Official Gazette, No. 23, https://qbz.gov.al/eli/fz/2018/23/c74feb8f-4bfa-4292-8b99-889a192b7ccc

Council of Ministers (2018c), *Decision No. 246 "On the Approval of the National Plan for European Integration 2018–2020"*. Official Gazette, No. 69, https://qbz.gov.al/eli/fz/2018/69/9e872d30-60ba-4535-8455-7f246648e6eb

Council of Ministers (2017), *Decision No. 141 "On the Organization and Functioning of the National Authority for Electronic Certification and Cyber Security"*. Official Gazette, No. 39, https://qbz.gov.al/eli/fz/2017/39/12b91bc2-be66-4285-8812-d43e675e9a78

Council of Ministers (2016a), *Decision No. 881 "On the Approval of the General National Plan of the Territory"*. Official Gazette No. 248, https://qbz.gov.al/eli/fz/2016/248/9c6f3829-d12f-49c4-b7a0-4c5fea9028be

Council of Ministers (2016b), *Decision No. 811 "On the Approval of the Transport Sector Strategy and Action Plan 2016–2020"*. Official Gazette, No. 220, https://qbz.gov.al/eli/fz/2016/230/1da489d0-011d-44ff-a6c0-803da4f3bc26

Council of Ministers (2015a), *Decision No. 410 "On the Establishment, Organization and Functioning of the State Technical and Industrial Inspectorate"*. Official Gazette, No. 137, https://qbz.gov.al/eli/fz/2015/88/174addd9-7064-45f1-ae82-072dfab72943

Council of Ministers (2015b), *Decision No. 893 "On the Organization and Functioning of the National Inspectorate of Territorial Protection"*. Official Gazette, No. 193, https://qbz.gov.al/eli/fz/2015/193/d4f35f2e-74e0-4d81-8ace-eb4e63fc17bf

Council of Ministers (2015c), *Decision No. 965, "On Inter-Institutional Cooperation of Management Structures, in Cases of Civil Emergencies and Crises"*. Official Gazette, No. 210, https://qbz.gov.al/eli/fz/2015/210/24ead646-ab0a-4f52-9aef-2e86bdd58799

Council of Ministers (2015d), *Decision No. 973, "On the Approval of the Cyber Security Policy Document 2015–2017"*. Official Gazette, No. 212, https://qbz.gov.al/eli/fz/2015/212/25adb531-b10d-4cd6-bcee-b89aac6afa77

Council of Ministers (2013), *Decision No. 663 "On the Approval of the Intersectoral Strategy of the Fight Against Organized Crime, Illegal Trafficking and Terrorism, 2013–2020 and the Action Plan for the Years 2013–2016"*. Official Gazette, No. 142, https://qbz.gov.al/eli/fz/2013/142/d7a99f16-6f94-4db0-98e3-1e4364037fe6

Council of Ministers (2002), *Decision No. 288, "On Determining and Taking Measures Against Fire and Rescue in Facilities of Economic and State Importance"*. Official Gazette, No. 37, https://qbz.gov.al/eli/fz/2002/37/13a629c1-689c-4bad-872e-0dc72f8fa1ff

Energy Regulatory Entity (2020), *Decision No. 126, "On the Approval of the Regulation on Cyber Security of Critical Infrastructures in the Electricity Sector"*, https://ere.gov.al/doc/Vendim_nr.126_2020(2).pdf

European Parliament and of the Council of the European Union (2016), *Directive 2016/1148 of the European Parliament and of the Council Concerning Measures for a High Common Level of Security of Network and Information Systems Across the Union*, EUR-Lex, Official Journal, https://eur-lex.europa.eu/legal-content/EN/TXT/PDF/?uri=CELEX:32016L1148&from=EN

European Parliament and of the Council of the European Union (2004), *Directive 2004/18/EC of the European Parliament and of the Council of 31 March 2004 on the Coordination of Procurement Procedures for the Awarding of Public Contracts for Works, Supplies and Services*, EUR-Lex, Official Journal, https://eur-lex.europa.eu/legal-content/EN/TXT/HTML/?uri=CELEX:32004L0018&from=en

NATO Science for Pace and Security Series (2015), Information and Communication Security. In Alessandro Niglia (Ed.), *The Protection of Critical Energy Infrastructure Against Emerging Security Challenges* (Vol. 43), IOS Press

TENTH CHAPTER (Kosova)

Assembly of the Republic of Kosovo (2018), *Law on Critical Infrastructure*, https://gzk.rks-gov.net/ActDetail.aspx?ActID=16313

Assembly of the Republic of Kosovo (2015), *Stabilization and Association Agreement Between the Republic of Kosovo and the European Union and the European Atomic Energy Community*, https://gzk.rks-gov.net/ActDetail.aspx?ActID=11240

Assembly of the Republic of Kosovo (2011), *Law for Protection Against Natural and Other Disasters*, https://gzk.rks-gov.net/ActDetail.aspx?ActID=2775

European Commission (2018), *Joint Action Plan on Counter-Terrorism for the Western Balkans*, https://ec.europa.eu/home-affairs/sites/homeaffairs/files/news/docs/20181005_joint-action-plan-counter-terrorism-western-balkans.pdf

Government of the Republic of Kosova (2019), *Concept Document on Information Systems and Networks Security Measures*, https://kryeministri-ks.net/wp-content/uploads/2019/10/Koncept-Dokumenti-per-Masat-e-Sigurise-se-Rrjeteve-te-Sistemeve-te-Informacionit-ang..pdf

Government of the Republic of Kosova (2017), *National Strategy Against Terrorism and Action Plan 2018–2022*, http://konsultimet.rks-gov.net/Storage/Consultations/09-35-08-10112017/2%20%20anglisht-National%20Strategy%20Against%20Terrorism%20and%20Action%20Plan%202018–2022_12%20October_without%20action%20plan.doc

Government of the Republic of Kosova (2015), *National Cyber Security Strategy and Action Plan 2016–2019*, http://www.kryeministri-ks.net/repository/docs/National_Cyber_Security_Strategy_and_Action_Plan_2016-2019_per_publikim_1202.pdf?__cf_chl_jschl_tk__=fcff8cab4ad9522370dcc26e57b156a57521d20e-1620219185-0-AZ-ogEylARj9teQOhsbz_A1p6w4cZKdKi23AkCniDWzlGwBpVPkYXhiwgtaFRheBI78TDrwOaBWrGjOGj3DO3He-L9setjbbCfBPDWH208JD3R5kG3mxOlK_G6LsaiD8flTnzLcAx_I5deveScuPoDTrbiVcxxn1PppAr1Q357i68BO0F7DBM7r3260AtjRgIxvx-tz9X8qGhJGJltuxix39oD9U4–2FTAXre2TJoEh76H9VqnynCGU_fKHhk_CoasJiEbut0zqfEgQZLcK-9tdVvbt5RtQg7hH0DNWlEbzFDt7nTMQYw3akILkF2co2VDaI1GXp5MfysAf_H5HlC8KtAGvxxvwUBhw-1mgWOQi8pY8F38X_kFHZiNovABfyoh8uj6UaOdYTbcDNz6BcBHuzrG20r6bVSMafuPyB-s2cnThrFW3DBZpV5oElU0faQ6wlXtkjyeOSIm6uP3vwJRe4tGXB-ysMKmLwzscKsU-x1pSWbltOURCs6HiDkKLabAQPbjmc0Hz2kUuZ9eNvoOU

Government of the Republic of Kosova (2012), *Regulation on the Methodology for Risk Assesment Development*, https://gzk.rks-gov.net/ActDetail.aspx?ActID=10293

Government of the Republic of Kosova (2010b), *National Response Plan*, https://mpb.rks-gov.net/Uploads/Documents/Pdf/EN/56/National%20Response%20Plan.pdf

ELEVENTH CHAPTER (Ukraine)

Biriukov, Dmytro; Kondratov, Sergiy; Nasvit, Oleg; Sukhodolia, Oleksandr (2015), *Green Paper for the Protection of Critical Infrastructure in Ukraine. Analytical Report*. Kyiv: NISS, https://www.researchgate.net/publication/299859217_Green_Paper_on_Critical_Infrastructure_Protection_in_Ukraine, Accessed: 10 September 2019

Cabinet of Ministers of Ukraine (2020a), *Order of the Cabinet of Ministers of Ukraine, 2020 № 943 "On Some Issues Regarding Critical Information Infrastructure Objects"*, https://zakon.rada.gov.ua/laws/show/943-2020-%D0%BF#Text (Ukrainian), Accessed: 19 November 2020

Cabinet of Ministers of Ukraine (2020b), *Order of the Cabinet of Ministers of Ukraine 2020 № 1109 "On Some Issues Regarding Critical Infrastructure Objects"*, https://zakon.rada.gov.ua/laws/show/1109-2020-%D0%BF#Text (Ukrainian), Accessed: 17 November 2020

Cabinet of Ministers of Ukraine (2019a), *The Draft of Law of Ukraine "On Critical Infrastructure and Its Protection"*, http://www.critical-protection.cf/doc/zakon_en.pdf (Unofficial translation), Accessed: 17 September 2019

Cabinet of Ministers of Ukraine (2019b), *Order of the Cabinet of Ministers of Ukraine 2019 No. 518 "On Approval of the General Requirements for Cyber Security of Critical Infrastructure Facilities"*, https://zakon.rada.gov.ua/laws/show/518-2019-%D0%BF (Ukrainian), Accessed: 17 September 2019

Cabinet of Ministers of Ukraine (2019c), *Brussels Hosts Trilateral Ukraine-EU-Russia Talks on Gas Transit*, https://www.kmu.gov.ua/en/news/u-bryusseli-vidbulisya-tristoronni-gazovi-peregovori-u-formati-ukrayina-yevropejska-komisiya-rf, (In Ukrainian), Accessed: 10 September 2019

Cabinet of Ministers of Ukraine (2017), *Resolution of the Cabinet of Ministers of Ukraine No. 1009-r "On Approval of the Concept for Building a State Critical Infrastructure Protection System"*, https://zakon.rada.gov.ua/laws/show/1009-2017-%D1%80, (In Ukrainian), Accessed: 10 September 2019

Council of the European Union (2008), *Council Directive 2008/114/EC of 8 December 2008 "On the Identification and Designation of European Critical Infrastructures and the Assessment of the Need to Improve Their Protection"*, EUR-Lex, Official Journal, https://eur-lex.europa.eu/legal-content/EN/TXT/?uri=celex:32008L0114

European Atomic Energy Community, European Union, The Member States, Ukraine (2014), *Association Agreement Between the European Union and Its Member States, of the One Part, and Ukraine, of the Other Part*, EUR-Lex, Official Journal, https://eur-lex.europa.eu/legal-content/EN/ALL/?uri=CELEX%3A22014A0529%2801%29

European Parliament and the Council of the European Union (2017), *Regulation (EU) 2017/1938 of the European Parliament and of the Council of 25 October 2017 Concerning Measures to Safeguard the Security of Gas Supply and Repealing Regulation (EU) No 994/2010*, EUR-Lex, Official Journal, https://eur-lex.europa.eu/eli/reg/2017/1938/oj

European Parliament and the Council of the European Union (2016), *Directive (EU) 2016/1148 of the European Parliament and of the Council of 6 July 2016 Concerning Measures for a High Common Level of Security of Network and Information Systems Across the Union*, EUR-Lex, Official Journal, https://eur-lex.europa.eu/eli/dir/2016/1148/oj

Horbulin, Volodymyr (2017), *The World Hybrid War: Ukrainian Forefront*, Kyiv: NISS

MECI (2018), *The Oder of the Ministry of Energy and Coal Industry of Ukraine № 448 "On Establishing of the Rules of Security of Electricity Supply"*, https://zakon.rada.gov.ua/laws/show/z1076-18 (In Ukrainian), Accessed: 10 September 2019

MECI (2015a), *The Oder of the Ministry of Energy and Coal Industry of Ukraine № 686 "On Establishing of the Rules of Security of Gas Supply,"* https://zakon.rada.gov.ua/laws/show/en/z1489-15/paran16#n16 (Ukrainian), Accessed: 10 September 2019

MECI (2015b), *The Oder of the Ministry of Energy and Coal Industry of Ukraine № 687 "On Establishing of the National Action Plan"*, http://zakon5.rada.gov.ua/laws/show/en/z1458-15/paran16#n16 (Ukrainian), Accessed: 10 September 2019

NISS (2017), Command and Staff Exercises on Protection of Critical Power Infrastructure and Response to the Threat of Failure of Its Stability at the National Level - COHERENT RESILIENCE 2017 (CORE 17), https://niss.gov.ua/en/node/2316, (In Ukrainian), Accessed: 10 August 2020

Parliament of Ukraine (2017), *The Law of Ukraine "On the Basic Principles of Ensuring the Cyber Security of Ukraine"*, https://zakon.rada.gov.ua/laws/show/2163-19; http://www.dsszzi.gov.ua/dsszzi/control/en/publish/article;jsessionid=2418BE1753A0CEC99666978C6B64648B.app2?art_id=289958&cat_id=35317 (Ukrainian), Accessed: 10 September 2019

Potii, Oleksandr; Tsyplinskyi, Yurii; Illiashenko, Oleg; Kharchenko, VVyacheslav (2020), Criticality Assessment of Critical Information Infrastructure Objects: A Category Based Methodology and Ukrainian Experience. In Dziech, A., Mees, W., Czyżewski, A. (Eds.), *Multimedia Communications, Services and Security* (Vol. 1284). MCSS 2020. Communications in Computer and Information Science. Cham: Springer, https://doi.org/10.1007/978-3-030-59000-0_7

President of Ukraine (2017), *Decree of the President of Ukraine №8/2017 "On Improvement of the Measures to Ensure Protection of Critical Infrastructure Objects"*, https://zakon.rada.gov.ua/laws/show/n0014525-16 (Ukrainian), Accessed: 10 September 2019

President of Ukraine (2016a), *Decree of the President of Ukraine №96/2016 "On Cyber Security Strategy of Ukraine"*, http://www.president.gov.ua/documents/962016–19836 (Ukrainian), Accessed: 10 September 2019

President of Ukraine (2016b), *Decree of the President of Ukraine № 92/2016 That Enacts the Decision of the National Security and Defense Council of Ukraine. Concept for the Development of the Defense and Security Sector*, http://zakon2.rada.gov.ua/laws/show/92/2016 (Ukrainian), Accessed: 10 September 2019

President of Ukraine (2015), *Decree of the President of Ukraine №287/2015 "On National Security Strategy of Ukraine"*, http://zakon3.rada.gov.ua/laws/show/287/2015 (Ukrainian), Accessed: 10 September 2019

State Service of Special Communication and Information Protection of Ukraine (SSSCIP) (2020), *Draft of the Order "On the Procedure of Independent Auditing of Critical Infrastructure Objects"*, https://cip.gov.ua/ua/news/povidomlennya-pro-oprilyudnennya-proektu-postanovi-kabinetu-ministriv-ukrayini-deyaki-pitannya-provedennya-nezalezhnogo-auditu-informaciinoyi-bezpeki-na-ob-yektakh-kritichnoyi-infrastrukturi (Ukranian), Accessed: 17 November 2020

Sukhodolia, Oleksandr (2018), Training as a Tool of Fostering a CIP Concept Implementation: Results of a Table Top Exercise on Critical Energy Infrastructure Resilience. *Information & Security: An International Journal*, 40, no. 2 (2018), https://infosec-journal.com/article/training-tool-fostering-cip-concept-implementation-results-table-top-exercise-critical, Accessed: 10 August 2020

Sukhodolia, Oleksandr, et al. (2019), *Organizational and Legal Aspects of Critical Infrastructure Security and Resilience*. Kyiv: NISS, https://niss.gov.ua/sites/default/files/2019-05/Dopov_Suchodolya_print.pdf(Ukrainian), Accessed: 17 September 2019

Sukhodolia, Oleksandr, et al. (2017), *Developing the Critical Infrastructure Protection System in Ukraine*. Kyiv: NISS, http://old2.niss.gov.ua/content/articles/files/niss_Engl_find-ruk-0e9af.pdf, Accessed: 10 September 2019

TWELFT CHAPTER (Belarus)

Balmaceda, Margarita M. (2006), *Belarus: Oil, Gas, Transit Pipelines and Russian Foreign Energy Policy*, London: GMB Publishing Ltd

Council of Europe (2020), *CyberEast*, https://www.coe.int/en/web/cybercrime/cybereast, Accessed: 28 June 2020

Council of Ministers of the Republic of Belarus (2014), *Resolution of the Council No. 508 "On Interministerial Infrastructure Coordinating Board*, http://www.economy.gov.by/uploads/files/G4P/Postanovlenie-Soveta-Ministrov-Respubliki-Belarus-ot-27-maja-2014-g.-508.docx, Accessed: 5 June 2020

European Commission (2019), *PPRD East 2 – Belarus*, https://europa.eu/capacity4dev/pprd-east-2/wiki/belarus, Accessed: 23 June 2020

Government of Belarus (2015a), *National Concept of Energy Security Till 2035*, http://www.government.by/ru/solutions/2337, Accessed: 5 June 2020

Government of Belarus (2015b), *The Strategy of Informatization Development in the Republic of Belarus for the Period 2016–2022 (2015)*, http://e-gov.by/zakony-i-dokumenty/strategiya-razvitiya-informatizacii-v-respublike-belarus-na-2016–2022-gody, Accessed: 1 June 2020

House of Representatives of Belarus (2016), *The Law No. 354-Z "On Industrial Safety"*, https://www.pravo.by/document/?guid=3961&p0=H11600354, Accessed: 1 June 2020

House of Representatives of Belarus (2015), *The Law No. 345-Z "On Public Private Partnerships"*, http://www.economy.gov.by/uploads/files/G4P/Zakon-Respubliki-Belarus-o-GChP-2.pdf, Accessed: 28 June 2020

House of Representatives of Belarus (2008a), *The Law No. 455-Z "On Information, Informatization and Data Protection"*, http://e-gov.by/zakony-ob-obrashheniyaxgrazhdan/ob-informacii-informatizacii-i-zashhite-informacii, Accessed: 1 June 2020

House of Representatives of Belarus (2008b), *The Law No. 418-Z "On Population Register"*, https://kodeksy-by.com/zakon_rb_o_registre_naseleniya.htm, Accessed: 18 November 2020

House of Representatives of Belarus (2004), *The Law No. 262-Z "On Technical Regulation and Standardization"*, http://law.by/document/?guid=3871&p0=H10400262e, Accessed: 1 May 2020

House of Representatives of Belarus (2002), *Law No. 77-Z "On the Fight Against Terrorism"*, http://kgb.by/ru/zakon77-3/, Accessed: 18 November 2020

House of Representatives of Belarus (2000), *The Law No. 363-Z "On Industrial Safety of Hazardous Production Facilities"*, http://base.spinform.ru/show_doc.fwx?rgn=1923, Accessed: 18 November 2020

House of Representatives of Belarus (1998), *The Law No. 141-Z "On the Protection of the Population and Territories from Natural and Technogenic Emergency Situations"*, https://mchs.gov.by/upload/iblock/1a9/1a91d57753f32e8bf96107e3bdc01be4.docx, Accessed: 1 June 2020

International Atomic Energy Agency (2018), *Peer Appraisal of Arrangements in the Republic of Belarus for Preparedness and Response for A Nuclear or Radiological Emergency*, https://www.google.com/url?sa=t&rct=j&q=&esrc=s&source=web&cd=&cad=rja&uact=8&ved=2ahUKEwj4-PmG7ZbtAhXssIsKHYVYCnEQFjAAegQIARAC&url=https%3A%2F%2Fwww.iaea.org%2Fsites%2Fdefault%2Ffiles%2Fdocuments%2Freview-missions%2Feprev-belarus-171018.pdf&usg=AOvVaw1reSqL6itf_vwg1ba4i-jEF, Accessed: 19 November 2020

The International Centre for Chemical Safety and Security (2019), *EU-CHEM-REACT 2. European Program on Exercises on Civil Protection in Belarus, Moldova and Ukraine,* https://www.iccss.eu/fileadmin/user-files/Documentation/Ukraine/EU_CHEM_REACT_2_short_Description_Nov.pdf, Accessed: 18 November 2020

Ministry of Architecture of the Republic of Belarus (2011), *Resolution No. 40 "On the Approval of the List of Critical Objects That Require Protection Against Terrorist Attacks"*, http://pravo.levonevsky.org/bazaby11/republic00/text934.htm, Accessed: 25 May 2020

The Official Internet Portal of the President of the Republic of Belarus (2020), *The President Approved the Lawmaking Plan for 2020* [Press Release], http://president.gov.by/ru/news_ru/view/utverzhden-plan-podgotovki-zakonoproektov-na-2020-god-22713/, Accessed: 5 June 2020

The Official Internet Portal of the President of the Republic of Belarus (2019), *Alexander Lukashenko Signed the Decree No. 449 "On Improving State Regulation in the Field of Information Protection"* [Press Release], http://president.gov.by/ru/news_ru/view/aleksandr-lukashenko-podpisal-ukaz-449-o-sovershenstvovanii-gosudarstvennogo-regulirovanija-v-oblasti-22566/, Accessed: 21 November 2020

OSCE (2013), *Presentation of the Good Practices Guide on Non-Nuclear Critical Energy Infrastructure Protection (NNCEIP) from Terrorist Attacks Focusing on Threats Emanating from Cyberspace,* https://www.osce.org/secretariat/107155, Accessed: 5 June 2020

Parliamentary Assembly of the Collective Security Treaty Organisation (2018), *Model Law "On Security of Critical Facilities"*, https://paodkb.org/documents/modelnyy-zakon-odkb-o-bezopasnosti-kriticheski-vazhnyh, Accessed: 10 June 2020

Parliamentary Assembly of the Collective Security Treaty Organisation (2016), *On the Strategy of CSTO Collective Security for the Period Till 2025*, https://en.odkb-csto.org/documents/documents/strategiya_kollektivnoy_bezopasnosti_organizatsii_dogovora_o_kollektivnoy_bezopasnosti_na_period_do_/, Accessed: 5 June 2020

Parliamentary Assembly of the Collective Security Treaty Organisation (2014), *Decision of the PA CSTO on Recommendations on Legislation Harmonization of the PA CSTO Member States on Critical Infrastructure Protection*, https://online.zakon.kz/Document//?doc_id=32202910#pos=0;0, Accessed: 10 June 2020

Parliamentary Assembly of the Collective Security Treaty Organisation (2002), *Charter of the Collective Security Treaty Organization*, https://paodkb.org/documents/ustav-organizatsii-dogovora-o-kollektivnoy-bezopasnosti, Accessed: 23 March 2020

Perevalov D. V. (2016), *Legal and Methodological Problems of Ensuring the Safety of Critical Facilities.*

President of Belarus (2019), *Presidential Decree of December 9, 2019 No. 449 "On the Enhancement of State Regulation in the Field of Information Security"*, http://president.gov.by/uploads/documents/2019/449uk.pdf, Accessed: 1 June 2020

President of Belarus (2015), *Presidential Decree No. 356 "About the Commission on Counteraction to Extremism and Fight Against Terrorism"*, https://www.google.com/url?sa=t&rct=j&q=&esrc=s&source=web&cd=&ved=2ahUKEwj0vfeC4qbqAhWdAhAIHWdQABgQFjAAegQIBBAB&url=https%3A%2F%2Fpravo.by%2Fupload%2Fdocs%2Fop%2FP31500356_1439931600.pdf&usg=AOvVaw28SjBhOvuuCQDMshcAm0IK, Accessed: 1 June 2020

President of Belarus (2011), *Presidential Decree No. 486 "On Some Measures for Safety of Critical Informatization Objects"*, https://oac.gov.by/public/content/files/files/law/decrees-rb/2011%20-%20486.pdf, Accessed: 1 June 2020

President of Belarus (2010), *Presidential Decree of November 9, 2010 No. 575 "On the Approval of the Security Concept of the Republic of Belarus"*, https://kgb.by/ru/ukaz575/, Accessed: 1 December 2020

Security Council of Belarus (2019), *Doctrine of Information Security of the Republic of Belarus*, http://president.gov.by/uploads/documents/2019/1post.pdf, Accessed: 1 June 2020

Sergi, Bruno S. (Ed.) (2019), *Modeling Economic Growth in Contemporary Belarus*. Bingley: Emerald Group Publishing

Yurtsov I. A. (2016), *Legal Regulation of Security of Especially Important State Facilities.*

Zoï Environment Network (2015), *Assessment "Environmental Safety of Main Pipelines in Belarus"*, https://zoinet.org/wp-content/uploads/2018/02/Assessment.22Oct2016EN.pdf, Accessed: 5 June 2020

THIRTEENTH CHAPTER (Moldova)

American Chamber in Moldova (2020), *Opinion on the Project of Law on the Protection of Infrastructure Objects Essential for Ensuring National Security and Public Order*, https://amcham.md/st_files/2020/06/19/072.%20Aviz%20AmCham%20-%20infrastructura%20esentiala.pdf

Cenusa, Denis (2019), *Moldova's Lasting Energy Dependency on Russia: The 'New' and 'Old' Traps*. Expert-Forum, https://expertforum.ro/en/kremlins-energy-policy-as-a-channel-of-influence-a-comparative-assessment/

Emerson, Michael; Cenusa, Denis (2018), *Deepening EU-Moldova Relations: What, Why and How? Second Edition*. Lanham: Rowman and Littlefield International

Energy Community (2020), *Implementation in Moldova*, https://energy-community.org/implementation/Moldova.html

European Commission (2018), *Action Document for EU4Digital: Improving Cyber Resilience in the Eastern Partnership Countries*, https://ec.europa.eu/neighbourhood-enlargement/sites/near/files/c_2018_8184_f1_annex_en_v1_p1_1000418.pdf

European Council (2001), *Convention on Cybercrime. European Treaty Series No. 185*, https://www.coe.int/en/web/conventions/full-list/-/conventions/rms/0900001680081561

Euruopean Union (2019), *On Establishing a Framework for the Screening of Foreign Direct Investments into the Union. Regulation 2019/452/19.03.2019*. EUR-Lex, Official Journal, https://eur-lex.europa.eu/eli/reg/2019/452/oj

European Union (2016), *Joint Framework on Countering Hybrid Threats a European Union Response*, EUR-Lex, Official Journal, https://eur-lex.europa.eu/legal-content/EN/TXT/?uri=CELEX%3A52016JC0018

European Union (2014), *EU-Moldova Association Agreement*, EUR-Lex, Official Journal, https://eur-lex.europa.eu/legal-content/EN/TXT/?uri=uriserv%3AOJ.L_.2014.260.01.0004.01.ENG

Information and Security Service (2018), *The Model Anti-Terrorist Passport*, Order No. 50/2018. Official Gazette, http://lex.justice.md/md/377828/

Ministry of Economy and Infrastructure of the Republic of Moldova (2019), *The Action Plan of the Ministry of Economy and Infrastructure for 2019*. Decree No. 48/2019, https://mei.gov.md/sites/default/files/raport_anual_de_activitate_al_ministerului_economiei_si_infrastructurii_pentru_anul_2019_.pdf

Moldovan Parliament (2020), *Draft Bill on the Protection of Essential Infrastructure Objects for Ensuring National Security and Public Order*, Draft Bill No. 163/2020, http://parlament.md/ProcesulLegislativ/Proiectedeactelegislative/tabid/61/LegislativId/5019/language/ro-RO/Default.aspx

Moldovan Parliament (2018a), *National Defense Strategy and Implementation Action Plan for the Period 2018–2022*, The Decision of the Parliament Nr. 134/2018. Official Gazette, https://www.legis.md/cautare/getResults?doc_id=110013&lang=ro

Moldovan Parliament (2018b), *Information Security Strategy of the Republic of Moldova for the Years 2019–2024*, Government Decision 257/2018. Official Gazette, https://www.legis.md/cautare/getResults?doc_id=111979&lang=ro

Moldovan Parliament (2017), *Preventing and Combating Terrorism*, Law No. 120/2017. Official Gazette, https://www.legis.md/cautare/getResults?doc_id=110295&lang=ro

Moldovan Parliament (2016), *National Defense Strategy of Moldova*, https://www.army.md/img/userfiles/doconline/sna.pdf

Moldovan Parliament (2008), *Concept of National Security*. Law No. 112/2008. Official Gazette, https://www.legis.md/cautare/getResults?doc_id=24400&lang=ro

Moldovan Parliament (2007), *Administration and Denationalization of Public Property*, Law No. 121/2007. Official Gazette, https://www.legis.md/cautare/getResults?doc_id=115473&lang=ro

Moldovan Parliament (2004a), *Investments in the Entrepreneurial Activity*, Law No. 81/2004. Official Gazette, https://www.legis.md/cautare/getResults?doc_id=100114&lang=ro

Moldovan Parliament (2004b), *State of Emergency, Siege and War*. Law No. 212/24.06.2004. Official Gazette, http://lex.justice.md/index.php?action=view&view=doc&lang=1&id=312211

Moldovan Parliament (2003), *National Defense*, Law No. 345/2003. Official Gazette, http://lex.justice.md/index.php?action=view&view=doc&lang=1&id=312812

Moldovan Parliament (2002), *Requisitions of Goods and Provision of Services in the Public Interest*, Law No. 1384/2002. Official Gazette, http://lex.justice.md/index.php?action=view&view=doc&lang=1&id=312965

Moldovan Parliament (1995), *The List of Units Whose Lands Destined for Agriculture Remain in State Ownership* (updated in 2019), Law No. 668/1995. Official Gazette, https://www.legis.md/cautare/getResults?doc_id=87315&lang=ro

Moldovan Government (2018a), *Regulation on the Protection of Critical Infrastructure Against Terrorism*, Government Decision No. 701/2018. Official Gazette, https://www.legis.md/cautare/getResults?doc_id=108835&lang=ro

Moldovan Government (2018b), *Regulation on Organizing and Conducting Anti-Terrorist Tests*, Government Decision No. 996/2018. Official Gazette, http://lex.justice.md/index.php?action=view&view=doc&lang=1&id=377719

Moldovan Government (2017), Mandatory Minimum Requirements for Cyber Security, Government Decision No. 201, 28.03.2017, https://mei.gov.md/sites/default/files/hg_201_2017_cerinte_minime_obligatorii_de_securitate_cibernetica.pdf

Moldovan Government (2015), *National Cyber Security Program*. Government Decision No. 811/2015. Official Gazette, https://gov.md/sites/default/files/document/attachments/intr53_0.pdf
Moldovan Government (2010), *Action Plan NATO-Moldova 2017–2019*, Government Decision nr. 746/2010. Official Gazette, https://www.legis.md/cautare/getResults?doc_id=21237&lang=ro
NATO (2020), *The NATO Science for Peace and Security Programme in Moldova*, https://www.nato.int/science/country-fliers/Moldova.pdf

FOURTEENTH CHAPTER (Georgia)

Government of Georgia (2015), *Decree №584 the on "The List of Subjects with High Risk for State Security"*, Legislative Herald of Georgia 11/2015, https://www.matsne.gov.ge/ka/document/view/3056223?publication=2
Government of Georgia (2014a), *Decree №312 on the Approval of "The List of Critical Information System Subjects"*, Legislative Herald of Georgia, https://www.matsne.gov.ge/ka/document/view/2333175?publication=1
Government of Georgia (2014b), *Decree №567 on Approval of "The List of Critical Information System Subjects in Defense Field"*, Legislative Herald of Georgia 09/2014, https://www.matsne.gov.ge/ka/document/view/2521602?publication=0
Government of Georgia (2013), *Order №427 on "The Approval of the List of Persons and Items of Protection of Special State Protection Service"*, Legislative Herald of Georgia 09/2013, https://www.matsne.gov.ge/ka/document/view/2019204?publication=0
Government of Georgia (2008), *Decree №93 the on "Property of the Special Importance"*, Legislative Herald of Georgia 55, https://www.matsne.gov.ge/ka/document/view/5590?publication=0
Minister of Economic Development of Georgia (2010), *Order №1–1/595 on "The List of the Property of Special Importance in the Civil Aviation Sector of Georgia"*, Legislative Herald of Georgia 46, https://www.matsne.gov.ge/ka/document/view/1011790?publication=1
Minister of Urbanization and Construction (2003), *Decree №58 on "The List of Objects of Strategic and Special Importance"*, Legislative Herald of Georgia 122, https://www.matsne.gov.ge/ka/document/view/1354315?publication=1
Parliament of Georgia (2018), *Law of Georgia on "Public-Private Partnership"*, Legislative Herald of Georgia 2273-II, https://matsne.gov.ge/en/document/download/4193442/0/en/pdf
Parliament of Georgia (2012), *Law of Georgia on "Information Security"*, Legislative Herald of Georgia 06/2012, https://www.matsne.gov.ge/en/document/view/1679424?publication=3
Parliament of Georgia (2010), *Law of Georgia on "State Owned Property"*, Legislative Herald of Georgia 48, https://www.matsne.gov.ge/en/document/view/112588?publication=29
Parliament of Georgia (2005), *Law of Georgia on "Licenses and Permits"*, Legislative Herald of Georgia 40, https://www.matsne.gov.ge/document/view/26824?publication=82
Parliament of Georgia (2003), *Law of Georgia on "Engineering-Geodesic Control and Objects of Strategic and Special Importance"*, Legislative Herald of Georgia 32, https://www.matsne.gov.ge/ka/document/view/13174?publication=3

Parliament of Georgia (1997), *Law on "Privatization of State Property, Privatization of Property of Local Self-Government Unit and Transfer of Right to Use"*, Parliamentary Gazzette, 29–30, 07/1997, https://www.matsne.gov.ge/ka/document/view/29920?publication=35

Parliament of Georgia (1996a), *Law of Georgia on "State Secrets"*, Legislative Herald of Georgia 10/1996, https://www.matsne.gov.ge/en/document/view/33280?publication=24

Parliament of Georgia (1996b), *Law of Georgia on "Promotion and Guarantees of Investment Activity"*, Parliamentary Gazzette, 29–30/5, https://www.matsne.gov.ge/en/document/view/33304?publication=15

FIFTEENTH CHAPTER (Armenia)

CERT-AM (2018), *Description of CERT-AM in Accordance with RFC 2350*, https://www.cert.am/rfc2350-CERT-AM.pdf

Council of Europe (2018), *Cybercrime and Cybersecurity Strategies in the Eastern Partnership Region, Updated Report 2018*, https://rm.coe.int/eap-cybercrime-and-cybersecurity-strategies/168093b89c

Council of Europe (2017), *Cybercrime Strategies, Procedural Powers and Specialised Institutions in the Eastern Partnership Region – State of Play*, https://rm.coe.int/3271-3608-report-on-eap-state-may2017/1680728bca

Delegation of the European Union to Armenia (2017), *Armenia and the EU*, https://eeas.europa.eu/delegations/armenia/896/node/896_az

EU Neighbours Portal (2020), *Facts and Figures About EU-Armenia Relations*, https://www.euneighbours.eu/en/east/stay-informed/publications/facts-and-figures-about-eu-armenia-relations-0

European Communities and Their Member States, and the Republic of Armenia (1999), *Partnership and Cooperation Agreement*, Official Journal of the European Communities, L 239/3, https://eeas.europa.eu/sites/eeas/files/eu-armenia_partnership_and_cooperation_agreement_en.pdf

European Union, European Atomic Energy Community, Republic of Armenia (2017), *EU-Armenia Comprehensive and Enhanced Partnership Agreement*, https://www.mfa.am/filemanager/eu/CEPA_ENG_1.pdf

Government of the Republic of Armenia (2020), *The Reform Strategy of the Police of the Republic of Armenia and the Action Plan for 2020–2022*, http://www.irtek.am/views/act.aspx?aid=104896

Government of the Republic of Armenia (2010), *Seismic Risk Reduction Program Derived from the Concept of the Development of the Seismic Safety System*, https://www.arlis.am/documentview.aspx?docID=65036

Government of the Republic of Armenia (2004), *The Lists of Buildings and Structures of State Bodies and Organizations Subject to State Protection Under the Government of the Republic of Armenia, as Well as the Most Important Objects*, https://www.arlis.am/documentview.aspx?docid=12702

National Security Council (2007), *National Security Strategy*, https://www.mfa.am/filemanager/Statics/Doctrineeng.pdf

Noravank Foundation (2018), *Critical Infrastructures and National Security*, http://www.noravank.am/upload/pdf/CRITICAL_STRUCTURES_book.pdf

President of the Republic of Armenia (2012), *Declaration of the President of RA on Approval National Strategy for Combating Terrorism in the Republic of Armenia*, https://www.arlis.am/documentview.aspx?docid=75353

Prime Minister of the Republic of Armenia (2020), *Decision About Composition and Operation Procedures of the Interagency Commission on Defense Reforms*, http://www.irtek.am/views/act.aspx?aid=104896

Prime Minister of the Republic of Armenia (2019), *Digitization Is a Key Quality Component in Public Services - Prime Minister Holds Meeting on Armenia's Digital Agenda*, http://www.irtek.am/views/act.aspx?aid=104896

Prime Minister of the Republic of Armenia (2018a), *The Charter and Structure of the National Security Service of the Republic of Armenia*, http://www.irtek.am/views/act.aspx?aid=95580

Prime Minister of the Republic of Armenia (2018b), *The Charter and Structure of the Ministry of Emergency Situations of the Republic of Armenia*, http://www.mes.am/files/docs/1433.pdf

SIXTEENTH CHAPTER (Azerbaijan)

ABADçılar əlverişli güzəştli kreditlərdən bəhrələnir (2018), https://abad.gov.az/post/36?locale=az

Aliyev, İ. (2020), *"Kritik İnfrastruktur" və onun təhdidlərdən qorunması: Dünya təcrübəsi və Azərbaycanda tətbiqinin zəruriliyi ("Critical Infrastructure" and Its Protection from Threats: World Experience and the Need for Its Application in Azerbaijan)*, https://1news.az/az/news/kritik-infrastruktur-ve-onun-tehdidlerden-qorunmasi-dunya-tecrubesi-ve-azerbaycanda-tetbiqinin-zeruriliyi

BP, *Azerbaijan Sustainability Report, 2011*, https://www.bp.com/content/dam/bp/country-sites/en_az/azerbaijan/home/pdfs/sustainability-reports/bp-sr-2011.pdf

BP, *Statistical Review of World Energy, 2011*, https://www.bp.com/en/global/corporate/energy-economics/statistical-review-of-world-energy.html

EU4Business (2019), *Investing in SMEs in the Eastern Partnership: Azerbaijan Country Report June 2019*, https://www.euneighbours.eu/sites/default/files/publications/2019-07/country_report_2019_azerbaijan.pdf

Focus Economics (2020), *Azerbaijan Economic Forecast*, https://www.focus-economics.com/countries/azerbaijan

GDSI Limited (2017), *Azerbaijan*, https://eapcivilsociety.eu/azerbaijan

The Igtisadiyyat (2018), *Azerbaijan*, https://azerbaijan.az/related-information/139

İmamverdiyev, Yadigar (2013), *Milli kibertəhlükəsizlik strategiyalarının analizi (Analysis of National Cyber Security Strategies)*, https://ict.az/uploads/konfrans/1_konfrans/03.pdf

Mission of the Republic of Azerbaijan to NATO (2020), *Emerging Security Challenges*, https://nato-pfp.mfa.gov.az/en/content/44

Montalbano, Elizabeth (2020), *PoetRAT Resurfaces in Attacks in Azerbaijan Amid Escalating Conflict*, https://threatpost.com/poetrat-resurfaces-azerbaijan-conflict/159917/

Muxtarli, Fuad (2018), *As Geopolitical Confrontation Deepens, South Caucasus Nations Need to Make Strategic Choices*, http://strati.az/mobile.php?id=1203

NATO *Parliamentary Assembly Document – Energy Security: Co-operating to Enhance the Protection of Critical Energy Infrastructures* (2008), https://www.nato-pa.int/document/2008-168-stcees-08-e-rev1-energy-security-tagarinski-report

Netblocks (2020), *Social Media Restricted in Azerbaijan Amid Clashes with Armenia Over Nagorno-Karabakh*, https://netblocks.org/reports/social-media-restricted-in-azerbaijan-amid-clashes-with-armenia-over-nagorno-karabakh-pA5LMj87

Nordea (2020), *Azerbaijan: Economic and Political Overview*, https://www.nordeatrade.com/en/explore-new-market/azerbaijan/economical-context

The State Statistical Committee of the Republic of Azerbaijan (2019), https://www.stat.gov.az/source/industry/?lang=en

UNCTAD (2018), The World Investment Report, https://unctad.org/topic/investment/world-investment-report

United States Congress (2001), *Uniting and Strengthening America Byproviding Appropriate Tools Required to Intercept and Obstruct Terrorism (USA Patriot Act) Act of 2001, Public Law 107–56 – OCT. 26, 2001*, pp. 1–132, https://www.govinfo.gov/content/pkg/PLAW-107publ56/pdf/PLAW-107publ56.pdf

World Economic Forum (2019), https://www.worldbank.org/en/publication/gfdr/data/global-financial-development-database

SEVENTEENTH CHAPTER (United States of America)

CEPCONSULT (2019), *The First Forum of the "EU-US Energy Council" in Brussels*, https://cepconsult.com/publications/the-first-forum-of-the-eu-us-energy-council-took-place-in-brussels/

Department of Defense (1998), *Critical Infrastructure Protection Plan*, https://fas.org/irp/offdocs/pdd/DOD-CIP-Plan.htm

Department of Homeland Security (2019), *Cyber Storm: Securing Cyber Space*, https://www.cisa.gov/cyber-storm-securing-cyber-space

Department of Homeland Security (2003), *Homeland Security Presidential Directive 7: Critical Infrastructure Identification, Prioritization, and Protection*, https://www.dhs.gov/homeland-security-presidential-directive-7#

Department of Homeland Security (2013a), *National Infrastructure Protection Plan (NIPP) 2013: Partnering for Critical Infrastructure Security and Resilience*, https://www.cisa.gov/sites/default/files/publications/national-infrastructure-protection-plan-2013-508.pdf

Department of Homeland Security (2013b), *National Infrastructure Protection Plan: Education and Awareness*, https://www.dhs.gov/xlibrary/assets/nipp_education.pdf

DHS, FEMA, DG ECHO (2011), *Administrative Arrangement on Cooperation in the Field of Civil Protection*, https://ec.europa.eu/echo/sites/echo-site/files/administrative_arrangement_us-fema.pdf

EU-U.S. Energy Council (2009), *2009 Energy Council Joint Press Statement*, https://ec.europa.eu/energy/sites/ener/files/documents/2009_energy_council_joint_press_statement.pdf

Federal Emergency Management Agency (2020), *Higher Education Project Listing of Emergency Management Programs by Type*, https://training.fema.gov/hiedu/collegelist/downloads/em%20programs%20by%20degree%20level.pdf?ver=20200205

Federal Emergency Management Agency (2014), *Emergency Management Higher Education Program Description*, https://training.fema.gov/HiEdu/brochure/2014%20-%20EM%20HiEd%20Pgm%20Description.doc
Government Accountability Office (2019), *Critical Infrastructure Protection: Actions Needed to Address Significant Cyber Security Risks Facing the Electric Grid*, GAO-19–332, https://www.gao.gov/assets/710/701079.pdf
Government Accountability Office (2006), *Critical Infrastructure Protection: Progress Coordinating Government and Private Sector Efforts by Sectors' Characteristics*, GAO-07–39, https://www.gao.gov/assets/260/252603.pdf
Minnesota Department of Public Safety (2020), *Emergency Management Certificate Programs*, https://dps.mn.gov/divisions/hsem/training/Pages/em-certification-program.aspx
Minnesota Department of Public Safety (2016), *Minnesota Critical Infrastructure Preparedness Planner Qualification Program*, https://dps.mn.gov/divisions/hsem/training/Documents/16-ci-planner-track-final.pdf
U.S. Embassy in Croatia (2020), *Croatia and U.S. Sign $5 Million Agreement to Launch Cyber Security Operations Center in Zagreb*, https://hr.usembassy.gov/croatia-and-u-s-sign-5-million-agreement-to-launch-cyber-security-operations-center-in-zagreb/?fbclid=IwAR0k0GmCS14ydqfCRfIJGGtWUivnCSMsxIjr3WUu8oNqJShx6CgV8F8h0Gs
White House (2013), *Presidential Policy Directive 21: Critical Infrastructure Security and Resilience*, https://fas.org/irp/offdocs/ppd/ppd-21.pdf
White House (1998), *Presidential Decision Directive/NSC-63: Critical Infrastructure Protection*, https://fas.org/irp/offdocs/pdd/pdd-63.pdf
White House (1996), *Executive Order 13010: Critical Infrastructure Protection*, https://fas.org/irp/offdocs/eo13010.htm

EIGHTEENTH CHAPTER (Canada)

Government of Canada (2018), *Annual Report on the State of the EU-Canada Relationship*, https://www.international.gc.ca/world-monde/international_relations-relations_internationales/can-eu_spa_report-aps_can-ue_rapport.aspx?lang=eng
Government of Canada (2016), *National Electric Grid Security and Resilience Action Plan*, https://www.nrcan.gc.ca/sites/www.nrcan.gc.ca/files/energy/pdf/Canadian%20Action%20Plan_EN.PDF
Government of Canada (2011), *Federal Emergency Response Plan*, https://www.publicsafety.gc.ca/cnt/rsrcs/pblctns/mrgnc-rspns-pln/mrgnc-rspns-pln-eng.pdf
Government of Canada (2004), *Position Paper on a National Strategy for Critical Infrastructure Protection*, http://www.acpa-ports.net/advocacy/pdfs/nscip_e.pdf
PESCO Secretariat (2019), *Cyber Rapid Response Teams and Mutual Assistance in Cyber Security*, https://pesco.europa.eu/project/cyber-rapid-response-teams-and-mutual-assistance-in-cyber-security/
Public Safety Canada (2019a), *Critical Infrastructure*, https://www.publicsafety.gc.ca/cnt/ntnl-scrt/crtcl-nfrstrctr/index-en.aspx
Public Safety Canada (2019b), *Emergency Management Strategy for Canada Toward a Resilient 2030*, https://www.publicsafety.gc.ca/cnt/rsrcs/pblctns/mrgncy-mngmnt-strtgy/mrgncy-mngmnt-strtgy-en.pdf

Public Safety Canada (2018a), *National Cross Sector Forum 2018–2020 Action Plan for Critical Infrastructure*, https://www.publicsafety.gc.ca/cnt/rsrcs/pblctns/pln-crtcl-nfr-strctr-2018-20/index-en.aspx

Public Safety Canada (2018b), *Evaluation of the Regional Resilience Assessment Program and the Virtual Risk Analysis Cell*, https://www.publicsafety.gc.ca/cnt/rsrcs/pblctns/2016-vltn-rrap/2016-vltn-rrap-en.pdf

Public Safety Canada (2018c), *Public Report on the Terrorist Threat to Canada*, https://www.publicsafety.gc.ca/cnt/rsrcs/pblctns/pblc-rprt-trrrsm-thrt-cnd-2018/pblc-rprt-trrrsm-thrt-cnd-2018-en.pdf

Public Safety Canada (2017), *An Emergency Management Framework for Canada – Third Edition*, https://www.publicsafety.gc.ca/cnt/rsrcs/pblctns/2017-mrgnc-mngmnt-frmwrk/2017-mrgnc-mngmnt-frmwrk-en.pdf

Public Safety Canada (2016), *Fundamentals of Cyber Security for Canada's Critical Infrastructure Community*, https://www.publicsafety.gc.ca/cnt/rsrcs/pblctns/2016-fndmntls-cybr-scrty-cmmnty/2016-fndmntls-cybr-Scrty-cmmnty-en.pdf

Public Safety Canada (2010a), *The National Strategy for Critical Infrastructure*, https://www.publicsafety.gc.ca/cnt/rsrcs/pblctns/srtg-crtcl-nfrstrctr/srtg-crtcl-nfrstrctr-eng.pdf

Public Safety Canada (2010b), *Action Plan for Critical Infrastructure*, https://www.publicsafety.gc.ca/cnt/rsrcs/pblctns/pln-crtcl-nfrstrctr/pln-crtcl-nfrstrctr-eng.pdf

Public Safety Canada and U.S. Department of Homeland Security (2010), *Canada-United States Action Plan for Critical Infrastructure*, https://www.cisa.gov/sites/default/files/publications/ip-canada-us-action-plan-2010-508.pdf

Index

For Product Safety Concerns and Information please contact our EU representative GPSR@taylorandfrancis.com
Taylor & Francis Verlag GmbH, Kaufingerstraße 24, 80331 München, Germany

www.ingramcontent.com/pod-product-compliance
Ingram Content Group UK Ltd.
Pitfield, Milton Keynes, MK11 3LW, UK
UKHW022216190926
475702UK00003B/386
9780367517182